W0018544

Mastering R

R is a statistical computing and graphics programming language that you can use to clean, analyze, and graph data. It is widely used by researchers from various disciplines to estimate and display results and by teachers of statistics and research methods. This book is a detailed guide for beginners to understand R with an explanation of core statistical and research ideas.

One of the powerful characteristics of R is that it is open-source, which means that anyone can access the underlying code used to run the program and add their own code for free. It will always be able to perform the latest statistical analyses as soon as anyone thinks of them. R corrects mistakes quickly and transparently and has put together a community of programming and statistical experts that you can turn to for help.

Mastering R: A Beginner's Guide not only explains how to program but also how to use R for visualization and modeling. The fundamental principles of R explained here are helpful to beginner and intermediate users interested in learning this highly technological and diverse language.

About the Series

The *Mastering Computer Science* covers a wide range of topics, spanning programming languages as well as modern-day technologies and frameworks. The series has a special focus on beginner-level content and is presented in an easy-to-understand manner, comprising:

- Crystal-clear text, spanning various topics sorted by relevance.

- Special focus on practical exercises, with numerous code samples and programs.

- A guided approach to programming, with step-by-step tutorials for absolute beginners.

- Keen emphasis on real-world utility of skills, thereby cutting the redundant and seldom-used concepts and focusing instead of industry-prevalent coding paradigm.

- A wide range of references and resources to help both beginner and intermediate-level developers gain the most out of the books.

Mastering Computer Science series of books start from the core concepts, and then quickly move on to industry-standard coding practices, to help learners gain efficient and crucial skills in as little time as possible. The books assume no prior knowledge of coding, so even the absolute newbie coders can benefit from this series.

Mastering Computer Science series is edited by Sufyan bin Uzayr, a writer and educator with over a decade of experience in the computing field.

For more information about this series, please visit: https://www.routledge.com/Mastering-Computer-Science/book-series/MCS

Mastering R
A Beginner's Guide

Edited by
Sufyan bin Uzayr

CRC Press
Taylor & Francis Group
Boca Raton London New York

CRC Press is an imprint of the
Taylor & Francis Group, an **informa** business

First Edition published 2024
by CRC Press
2385 NW Executive Center Drive, Suite 320, Boca Raton, FL 33431

and by CRC Press
2 Park Square, Milton Park, Abingdon, Oxon, OX14 4RN

CRC Press is an imprint of Taylor & Francis Group, LLC

© 2024 Sufyan bin Uzayr

Reasonable efforts have been made to publish reliable data and information, but the author and publisher cannot assume responsibility for the validity of all materials or the consequences of their use. The authors and publishers have attempted to trace the copyright holders of all material reproduced in this publication and apologize to copyright holders if permission to publish in this form has not been obtained. If any copyright material has not been acknowledged please write and let us know so we may rectify in any future reprint.

Except as permitted under U.S. Copyright Law, no part of this book may be reprinted, reproduced, transmitted, or utilized in any form by any electronic, mechanical, or other means, now known or hereafter invented, including photocopying, microfilming, and recording, or in any information storage or retrieval system, without written permission from the publishers.

For permission to photocopy or use material electronically from this work, access www.copyright.com or contact the Copyright Clearance Center, Inc. (CCC), 222 Rosewood Drive, Danvers, MA 01923, 978-750-8400. For works that are not available on CCC please contact mpkbookspermissions@tandf.co.uk

Trademark Notice: Product or corporate names may be trademarks or registered trademarks and are used only for identification and explanation without intent to infringe.

ISBN: 9781032415215 (hbk)
ISBN: 9781032415185 (pbk)
ISBN: 9781003358480 (ebk)

DOI: 10.1201/9781003358480

Typeset in Minion
by KnowledgeWorks Global Ltd.

For Mom
_____

Contents

About the Editor

Sufyan bin Uzayr is a writer, coder, and entrepreneur with over a decade of experience in the industry. He has authored several books in the past, pertaining to a diverse range of topics, ranging from History to Computers/IT.

Sufyan is the Director of Parakozm, a multinational IT company specializing in EdTech solutions. He also runs Zeba Academy, an online learning and teaching vertical with a focus on STEM fields.

Sufyan specializes in a wide variety of technologies, such as JavaScript, Dart, WordPress, Drupal, Linux, and Python. He holds multiple degrees, including ones in management, IT, literature, and political science.

Sufyan is a digital nomad, dividing his time between four countries. He has lived and taught in universities and educational institutions around the globe. Sufyan takes a keen interest in technology, politics, literature, history, and sports, and in his spare time, he enjoys teaching coding and English to young students.

Learn more at sufyanism.com

Acknowledgments

There are many people who deserve to be on this page for this book would not have come into existence without their support. That said, some names deserve a special mention, and I am genuinely grateful to:

- My parents, for everything they have done for me.

- The Parakozm team, especially Divya Sachdeva, Jaskiran Kaur, and Simran Rao, for offering great amounts of help and assistance during the book-writing process.

- The CRC team, especially Sean Connelly and Danielle Zarfati, for ensuring that the book's content, layout, formatting, and everything else remain perfect throughout.

- Reviewers of this book, for going through the manuscript and providing their insight and feedback.

- Typesetters, cover designers, printers, and everyone else, for their part in the development of this book.

- All the folks associated with Zeba Academy, either directly or indirectly, for their help and support.

- The programming community in general, and the web development community in particular, for all their hard work and efforts.

—**Sufyan bin Uzayr**

Zeba Academy – Mastering Computer Science

The "Mastering Computer Science" series of books are authored by the Zeba Academy team members, led by Sufyan bin Uzayr, consisting of:

- Divya Sachdeva

- Jaskiran Kaur

- Simran Rao

- Aruqqa Khateib

- Suleymen Fez

- Ibbi Yasmin

- Alexander Izbassar

Zeba Academy is an EdTech venture that develops courses and content for learners primarily in STEM fields, and offers educational consulting and mentorship to learners and educators worldwide.

Additionally, Zeba Academy is actively engaged in running IT Schools in the CIS countries, and is currently working in partnership with numerous universities and institutions.

For more info, please visit https://zeba.academy

## CHAPTER 1

Introduction to R

IN THIS CHAPTER

➤ Data Science

➤ Introduction to R

➤ History

➤ Applications

This is the first chapter of the R book. This chapter will provide you a brief knowledge of its fundamentals. The R is basically used for data science. Before deep diving into R, let's have a look at the fundamental of data science.

DATA SCIENCE

Data science continues to emerge as one of the most promising and in-demand career paths for qualified professionals.[1] Today, data professionals understand that they must overcome the traditional skills of big data analysis, data mining, and programming. In order to uncover useful insights for their organizations, data scientists must master the full spectrum of the data science lifecycle and have a level of flexibility and understanding to maximize return at each stage of the process.

Data science is an exciting discipline that allows you to transform raw data into understanding and knowledge. The goal of the "R for Data Science" is to help to learn the most important tools in R that will enable

DOI: 10.1201/9781003358480-1

to do data science. After reading this book, you will have tools to tackle a wide range of data science challenges using the best parts of R.

Data Science Life Cycle

A data science lifecycle is an iterative set of data science steps that you take to deliver a project or analysis. Because every data science project and team is different, every particular data science lifecycle is different. However, most data science projects tend to go through the same general lifecycle of data science steps.

Some data science lifecycles focus narrowly on only the data, modeling, and evaluation steps.[2] Others are more complex and start with business understanding and end with deployment.

And, the one we're going to go over is even more extensive to include surgery. It also emphasizes agility more than other lifecycles.

This life cycle has five steps:

- Problem definition

- Data investigation and cleaning

- Minimum viable model

- Deployment and enhancements

- Data science ops

These are not linear steps of data science. You start with step one and then continue with step two. However, from there you should flow naturally between the steps as needed; few small iterative steps are better than a few larger complex phases. Note that the lifecycle focuses on project steps. For a more comprehensive view that includes a collaborative framework, review the data science process.

Data Science Process

The data science process can make or break a team. Indeed, we see time and time again that many of the reasons data science projects fail are not technical in nature but rather stem from process-related issues. Just throwing computing power and PhDs at the problem doesn't work. Rather, the right mix of people, data, technology, and processes are key ingredients in the recipe for a successful execution of data science projects.

INTRODUCTION TO R

R is a language and environment for statistical computing and graphics.[3] It is a GNU project same as the S language and environment that was developed at Bell Laboratories formerly AT and T but now Lucent Technologies by John Chambers and colleagues. It can be thought as a different implementation of S. There are some important differences, but code written for S runs unchanged under R.

R provides a variety of statistical (linear and nonlinear modeling, classical statistical tests, time series analysis, classification, clustering, and many more) and graphical techniques and is highly extensible. The S language is often the medium of choice for research in statistical methodology, and the R language provides a path to participate in this open-source activity.

One of its strengths is the ease with that its well-designed, publication-quality charts can be created, including mathematical symbols and formulas where needed. Great care has been taken with the default settings for minor design choices in the graphics, but the user gets full control.

R is available as a free-of-cost software under the terms of the GNU General Public License from the Free Software Foundation in source code form. It compiles and runs on a wide range of UNIX platforms and similar systems (including FreeBSD and Linux), Windows, and macOS.

ENVIRONMENT OF R

R is an integrated set of facilities software for data manipulation, computation, and graphical display. It includes efficient equipment for handling and storing data, a set of operators for calculations on arrays, especially matrices, a large, coherent, integrated collection of intermediary tools for data analysis, graphic devices for data analysis and display on screen or in printed form a well-developed, simple and efficient programming language that includes conditionals, loops, user-defined recursive functions, and input and output devices. The term "environment" is intended to characterize it as a planned and coherent system, rather than an increment accretion of very specific and inflexible tools, as is often the case with other data analysis software.

R, like S, is designed based on a real computer language and allows users to add more functionality by defining new functions. Much of the system itself is written in the R dialect of the S language, making it easy for users to follow algorithmic choices made. For intensive tasks, C, C++, and

Fortran code can link and called at runtime. Advanced-user programmers write C program to manipulate R objects directly.

Many users think of R as a statistical system. We think of it as the environment in which statistical techniques are implemented. R can be (easily) extended via packages. About eight packages come with the R distribution, and many more are available through the Comprehensive R Archive Network (CRAN) family of websites, covering a very wide range of modern statistics. R has own LaTeX-like documentation format, which is used to provide comprehensive documentation, both online in many formats and in print.

ORIGIN OF R

R is an alternative implementation of a statistical programming language called S.[4] S-PLUS was developed after S as its commercial version. R was introduced later by Ross Ihaka and Robert Gentleman in 1991. Although R is independent of S-PLUS, much of its code works without any changes for R. The first version was released in 1995 as an open-source software package under GNU General Public License.

HISTORY

R is an open-source and free implementation of the S programming language combined with the lexical scope semantics from a scheme that allows objects to be defined in predefined blocks rather than throughout the code. S was created by Rick Becker, Doug Dunn, John Chambers, Jean McRae, and Judy Schilling at Bell Laboratories around 1976. The language was designed for statistical analysis and is an interpreted language whose code can be directly executed without a compiler. Many programs written for S run unchanged in R. As a dialect of Lisp, a scheme was created by Gerald J. Sussman and Guy L. Steele Jr. at MIT around 1975.

R was implemented in the early 1990s by Robert Gentleman and Ross Ihaka, both faculty members at the University of Auckland. The R language was closely modeled after the S language for statistical computing created by John Chambers, Rick Becker, Trevor Hastie, Allan Wilks, and others at Bell Labs in the mid-1970s and made publicly available in the early 1980s. Robert and Ross founded R as an open-source project in 1995. Since 1997, the R project has been managed by the R Core Group. And in February 2000, the first release of R was released. See also Ross Ihaka's brief account of how R began, highlighting some of the connections between R and S.

The R Core Team was founded in 1997 to further develop the language. As of January 2022, it consists of Chambers, Gentleman, Ihaka, and Mächler, plus statisticians Douglas Bates, Peter Dalgaard, Kurt Hornik, Michael Lawrence, Friedrich Leisch, Sebastian Meyer, Paul Murrell, Martyn Plummer, Brian Ripley, Uwe Ligges, Thomas Lumley, Deepayan Sarkar, Duncan Temple Lang, Luke Tierney, and Simon Urbanek as well as computer scientist Tomáš Kalibera. The members were Stefano Lacus, Guido Masarotto, Heiner Schwarte, Seth Falcon, Martin Morgan, and Duncan Murdoch. In April 2003, the R Foundation was established as a nonprofit organization to further support the R project.

R programming language is the higher version of the S language. John Chamber is the creator of the S language in 1976 at the Bell laboratories.

- In 1988, the version of the S programming language came into existence with the name S-PLUS. The R language is the unchanged version of S-PLUS.

- In 1991, the two gentlemen, Ross Ihaka and Robert Gentlemen found the alternative of S language that is independent on S-PLUS.

- In 1993, they started publicizing the alternative. It has named after the first two R authors and two S authors.

- In 1995, another gentleman Martin Maechler convinced the two to make R as free and open-source software with a number of other formers. The official version was released.

- On April 23, 1997, the CRAN commercially announced minor changes and contributed packages.

- On February 29, 2000, the official "stable beta" version was released.

- The latest released version 4.1.2 on November 01, 2021.

R VERSUS PYTHON

Python and R both are open-source software languages that have been around for a while. When comparing R versus Python, some feel that Python is a more general programming language. Python is often taught in introductory programming courses and is the primary language for many machine-learning workflows. R is typically used in statistical computing. RStudio notes that R is often taught in statistics and data science courses. He adds that many machine-learning interfaces are written in Python, while many statistical methods are written in R.

In terms of the R environment versus Python is an R environment ideal for manipulating data and graphs. Some applications of Python External link such as open_in_new include web development, numerical computation, and software development. Additionally, while R has many packages, Python has many libraries dedicated to data science.

Is R versus Python better or not may depend on what you use each of them for. Knowledge of both languages can be beneficial in data science. In fact, RStudio notes that many data science teams are "bilingual" using both R and Python.

Here are some important things to know about R in data science:

- R is open-source software. R is free and customizable because it is an open-source software. R's open interfaces enable integration with other applications and systems. Open-source software has a high-quality standardExternal link:open_in_new because more people use and iterate on it.

- R is a programming language. As a programming language, R provides objects, operators, and functions that allow users to explore, model, and visualize data.

- R is used for data analysis. R in data science is used to process, store, and analyze data. It can be used for data analysis and statistical modeling.

- R is an environment for statistical analysis. R has various statistical and graphical options. R FoundationExternal link:open_in_new states that it can be used for classification, clustering, statistical tests, and linear and nonlinear modeling.

- R is a community. R Project contributorsExternal link:open_in_new includes individuals who have suggested enhancements, reported bugs, and created supplemental packages. Although there are more than 20 official contributors, the R community extends to those who use the open-source software independently.

How Does R Work?

R is both a software and a language.[5] There are only a few things a user should be aware of to understand how R works, which we can separate between these two aspects.

R Is Software

- It should be installed on the computer before the lesson.

- It has a basic set of abilities that can be enhanced by adding packs.

- It can be used as a calculation engine by other software such as RStudio.

- It is mostly run using keyboard commands.

Usage

Not only in the IT companies using the R programming language but also providing services to small, medium, and large enterprises for their business intelligence.[6] These companies use R for machine-learning products because they can create products for processing statistical data and computing tools that can create services that manipulate data. Using R, the relevance of large data types can be performed and ultimately derive a meaningful summary from the data.

There are various companies that use R, such as

- IBM

- Wipro

- Accenture

- Infosys

- TCS

- Microsoft

- Google

Use of the R Programming Language in Various Areas

R is an open-source coding language used for data visualization, statistical analysis, and data science.[7] Since R is open source, it has huge communities that are constantly working to improve their environment and help members around the world to innovate and improve. R is highly flexible and compatible with different technologies. It consists of more than 10,000 packages and libraries to enhance and extend its significant capabilities. It also has libraries for both dynamic and static graphics. Now this blog will help you find out where R is used in different sectors, such as social media, research and academics, and much more. We have listed some uses of this programming language through which one can easily understand its importance. Let's take a look at the list below:

R in Research and Academics
Many students use R to perform various statistical analyses and calculations because it is a statistical research tool. Statistics techniques can be linear and nonlinear modeling, time series analysis, classification, classical statistical tests, clustering, and all other implementations of R and its libraries.

This language is used for deep learning as well as machine-learning research. Its library can be used to support both supervised and unsupervised learning. And, learn machine programming, it's a popular programming language.

With libraries that facilitate supervised and unsupervised learning, R is one of the most widely used languages for machine learning.

Other research involves large data sets, such as big data, searching for genetic anomalies and patterns, and various drug formulations, all use R to sift through a large collection of relevant data and draw meaningful conclusions from it.

R Use Cases in Research and Academia

1. Cornell University: Cornell encourages its researchers and students to use R for all their research involving statistical calculations.[8]

2. UCLA: The University of California, LA uses R to teach statistics and data analysis to its students.

R in IT Sectors
IT companies are not only using R for their own business intelligence, but also offering such services to other small, medium, and large enterprises. They also use it for their machine-learning products. They use R to build statistical tools and data processing products and to build other data manipulation services. Some large IT companies that use R, such as

1. Accenture

2. IBM

3. Infosys

4. Paytm

5. Tata Consulting Services

6. Wipro

Some real-world examples of R in the IT sector are given below:[9]

1. Mozilla: It uses R to visualize web activity for its Firefox browser.

2. Microsoft: It uses R as a statistical engine within Azure Machine Learning. They use it for the Xbox matchmaking service.

3. Foursquare: R works behind the scenes on Foursquare's recommendation engine.

4. Google: Google uses R to improve search results, to provide better search suggestions, to calculate the return on investment of its advertising campaigns, to increase the effectiveness of its online advertising, and to predict its economic activity.

Some of the important applications of the R programming language in data science are as follows.

R in Finance

Data Science is most prevalent in the financial industry.[10] R is the most famous tool for this role. It provides an advanced statistical suite capable of performing all the necessary financial tasks. Using the R, financial institutions are able to measure downside risk, adjust risk performance, and use visualizations such as candlestick charts, density charts, drawdown charts, and so on.

R provides tools for moving averages, autoregression, and time series analysis that form the core of financial applications. R is widely used for credit risk analysis in firms such as ANZ and portfolio management.

Financial industries also use R statistical processes with time series to model their stock market movements and forecast stock prices. R also provides financial data-mining facilities through its packages such as quantmod, prefetch, TFX, pwt, and so on. It makes it easy to extract data from online assets. With the help of shiny, you can also showcase your financial products through vivid and engaging visualizations.

R for Social Media

For many beginners in the Data Science and R, social media is a data playground. The sentiment analysis and other forms of social media data mining are the important statistical tools used with R.

Social media is a challenging field for Data Science because the data prevalent on social media websites is unstructured in nature. R is used for social media analysis to segment and target potential customers to sell your products.

In addition, user sentiment mining is another popular category in social media analytics. With the help of R, the companies are able to model statistical tools that analyze user sentiment, allowing them to improve their experience.

R in Banking

Banks make extensive use of the mortgage foreclosure model, which allows them to take over the property in case of default on the loan. Mortgage discount modeling includes sale price distribution, sale price volatility, and expected shortfall calculation. For these purposes, R is often used together with proprietary tools such as SAS. It is also used in conjunction with Hadoop to facilitate customer quality analysis, customer segmentation, and retention. Bank of America uses R for financial reporting. With the help of R, data scientists at BOA are able to analyze financial losses and use R's visualization tools.

There are various banking firms that use R for risk analysis and risk modeling. Various banks use R with their proprietary software, such as SAS. It is used for mortgage discount modeling, volatility modeling, client scoring, fraud detection, statistical modeling, loan stress test simulation, and much more. In addition to statistical analysis, it is used for data visualization, business intelligence, customer-quality calculation, customer segmentation, and customer retention.

R in Healthcare

Genetics, bioinformatics, drug discovery, and epidemiology are some of the fields in healthcare that make heavy use of R. With the help of R, these companies are able to process data and process information, which provides the basic background for further analysis and data processing.

For more advanced processing such as drug discovery, R is most widely used for conducting preclinical studies and analyzing drug safety data. It also provides its users with a suite for performing data analysis and live visualization tools.

R is also famous for its Bioconductor package, which provides various functions for analyzing genomic data. R is also used for the statistical modeling in the field of epidemiology, where the data scientists analyze and predict the spread of disease.

R in Manufacturing

Manufacturing companies like Ford, Modelez, and John Deere use R to analyze customer sentiment. This helps them to optimize their product according to the trends of consumer interests and also adjust the production volume to the changing market demand. They also use it to minimize their production costs and maximize profit.

R in the Governmental Department

R serves to keep records and process their census in the department of state administration. It helps them with effective administration and lawmaking. It is also used for services such as weather forecasting, drug regulation, and disaster impact analysis.

SOME OTHER APPLICATIONS OF R

Here are some other R applications you can use to make better decisions:

- R is primarily used for descriptive statistics. The descriptive statistics summarize the main features of the data. R is used for various purposes in summary statistics, such as central tendency, measuring variability, finding kurtosis, and skewness.

- R is most commonly used for exploratory data analysis. The most popular R package ggplot2 is considered one of the best visualization libraries due to its aesthetics and interactivity.

- R is used for analyzing both discrete and continuous probability distributions. For example, you can draw a Poisson distribution using the ppois() function. Similarly, you can use the dbinom() function to plot the binomial distribution.

- R also allows hypothesis testing to validate the statistical models.

- You can find the correlation between variables in R using the lm() function, which is used to determine both linear regression and multivariate linear regression.

- With R, you can take advantage of the tidyverse package, which is used for data organization and data preprocessing.

- R also provides a package of interactive web applications called shiny. With this package, you can develop attractive visualizations that can be embedded on your website.

- Additionally, with R you can develop predictive models that use machine-learning algorithms to find occurrences of future events.

- You can integrate R language with Hadoop and HDFS to quickly and efficiently process large datasets such as social media data.

- R is useful for developing statistical software packages and for implementing analytical processing in other software packages.

Various reasons to use R language:[11]

- R programming is used as a tool for machine learning, statistics, and data analysis. The objects, functions, and packages can be easily created with R.

- It is a platform-independent language. This means it can be used on all operating systems.

- It is a free and open-source language. This means that it can be installed by anyone in any organization without purchasing a license.

- The R programming language is not only a statistical package but also allows us to integrate with other languages (C, C++). So you can easily work with many data sources and statistical packages.

- The R programming language has a large community of users and is growing every day.

- R is currently one of the most sought-after programming languages in the Data Science job market, making it the hottest trend today.

- R can perform operations on vectors, it doesn't require too many loops.

- It can pull data from the APIs, servers, SPSS files, and many other formats.

- It is useful for web scraping.

- It can perform various complex mathematical operations with a single command.

- Using Markdown, it can create attractive reports that combine plain text with code and visualizations of the results.

- Due to the large number of researchers and statisticians who use it, the R community is often the first to see new ideas and technologies.

THE MOST POPULAR R PACKAGES

R packages are defined as collections of functions, sample data, documentation, and compiled code. These elements are kept in a directory called "library" in the R environment and are installed by default during installation.

R packages enhance the performance of R by enhancing existing features and collecting R feature sets into a single unit. In addition, the package is a reusable resource that makes a programmer's life much easier.

R PACKAGE

R packages are extensions of the R statistical programming language.[12] Its packages contain code, data, and documentation in a standardized collection format that can be installed by R users, typically through a centralized software repository such as CRAN. A large number of packages are available for R and their ease of installation and use have been cited as a major factor that has led to the language's widespread adoption in data science. R is distributed with 15 "core packages": base, compiler, datasets, grDevices, graphics, grid, methods, parallel, spline, statistics, stats4, tcltk, tools, translations, and utils.

Compared to libraries in another programming language, R packages must conform to a fairly strict specification. The Writing R Extensions manual specifies a standard directory structure for R source code, data, documentation, and package metadata, allowing them to be installed and loaded using R's built-in package management tools. Packages distributed on CRAN must conform to other standards. According to John Chambers, while these requirements "put significant demands" on package developers, they improve the usability and long-term stability of packages for end users.

In addition, there are 15 "recommended packages" from CRAN that are included in R binary distributions: KernSmooth, MASS, Matrix, boot, class, cluster, code tools, foreign, lattice, mgcv, nlme, nnet, rpart, space, and survival.

A group of packages called Tidyverse, which can be considered a "dialect of R," is becoming increasingly popular in the R ecosystem. As of 06/13/2020, Metacran listed seven of the eight core Tidyverse packages in the list of most downloaded R packages. The package group aims to provide a comprehensive collection of functions for solving common data science tasks, including data import, cleaning, transformation, and visualization (especially with the ggplot2 package). R Infrastructure packages support R package coding and development.

The capabilities of R are extended through user-created packages that offer statistical techniques, plotting facilities, import/export, reporting (RMarkdown, knitr, Sweave), and so on. These packages and their ease of installation and use have been cited as driving the language's widespread adoption in data science. The packaging system is used by researchers to organize research data, coding, and report files in a well-systematic way for sharing and archiving.

The basic installation includes several packages. Additional packages are available on CRAN, Bio-conductor, Omega hat, R-Forge, GitHub, and others. "Task Views" on the CRAN sites lists packages in areas including finance, genetics, high-performance computing, machine learning, medical imaging, meta-analysis, social science, and spatial statistics. R has been designated by the FDA as suitable for the interpretation of clinical research data. Microsoft maintains a daily CRAN snapshot that dates back to September 17, 2014.

Other sources of R packages include R-Forge, a platform for the collaborative development of R packages. The Bioconductor project provides genomic data analysis packages, including object-oriented data processing and data analysis tools from Affymetrix, cDNA microarray, and next-generation high-throughput sequencing methods.

A group of packages called Tidyverse, which can be considered a "dialect" of R, is becoming increasingly popular among developers. It seeks to provide a comprehensive collection of functions for solving common data science tasks, including data import, cleaning, transformation, and visualization (especially with the ggplot2 package). Dynamic and interactive graphics are only available through additional packages. R is one of the five languages with an Apache Spark API, along with Scala, Java, Python, and SQL.

COMPREHENSIVE R ARCHIVE NETWORK

The CRAN is a central R software repository supported by the R Foundation. It contains an archive of the latest versions of the R distribution, documentation, and added R packages. It includes both source packages and precompiled binaries for Windows and macOS. As of November 2020, there are over 16,000 packages available. CRAN was created by two, Kurt Hornik and Friedrich Leisch, in 1997, named in parallel with other early packaging systems such as TeX's CTAN (released 1992) and Perl's CPAN (released 1995). As of 2021, it is still maintained by Hornik and a team of volunteers. The main site is located at the Vienna University of Economics and Business and is mirrored on servers around the world.

The "Task Views" page on CRAN lists a wide range of tasks (in fields such as finance, genetics, high-performance computing, machine learning, medical imaging, meta-analysis, social sciences, and spatial statistics) for which R packages are available. Another method to browse CRAN packages is provided by Metacran, which also maintains lists of recommended, most downloaded, trending, or most dependent packages.

The number of CRAN packages are growing exponentially over the years, and as of 2018, an average of 21 new or updated packages were submitted each day. Since each post is manually reviewed by a small team of CRAN maintainers, many of whom are "nearing retirement age" according to R core developer Peter Dalgaard, there is concern that it is not sustainable in the long term. The growth of CRAN revealed the limitations of its dependency management infrastructure, notably the fact that it assumes that dependencies refer to the latest version of a package, which means that new releases of CRAN packages must be backward compatible and that CRAN packages cannot have non-CRAN dependencies. This has also led to concerns about declining package quality.

IDE AND EDITORS

R is a programming language mainly designed for statistical computing and data science.[13] R Programming is software supported by the R Foundation for statistical computing and nonprofit organizations.

Being a statistical software package, it has grown in popularity among researchers and data miners who use it for data-mining surveys and data analysis. Its source code was basically written in C, Fortran, and R.

R programming is freely available to the public under the GNU license. R has precompiled binaries for common operating systems. R can be run on the command line for terminal experts and graphical user interfaces (GUIs) in integrated development environments (IDEs).

RStudio

RStudio was developed by RStudio Inc. founded by JJ Allaire. It is available in two formats, one running locally as a desktop application known as Desktop and Server RStudio, which allows access to RStudio via a web browser while running remotely on a Linux server.

RStudio is available under the free GNU AGPL v3 license, an open-source license that guarantees the freedom to share code. RStudio is available in a prepackaged distribution for Windows, Linux, and macOS, while RStudio Server can run on Ubuntu, SLES, OpenSUSE, Debian, Linux, and CentOS.

R Tools for Visual Studio

As a powerful coding IDE, Visual Studio has brought an amazing experience to R programmers. Now you can enjoy IDE features while writing R programs with the newly released R Tools for Visual Studio (RTVS) product, which Microsoft released under the free and open-source MIT license.

RTVS is a plug-in for the Microsoft Visual Studio IDE used to provide support for programming in the R language. It supports IntelliSense, debugging, rendering, remote execution, SQL integration, and more. It is distributed as an open-source software under the Apache License 2.0 and is primarily developed by Microsoft. The version released was 0.3 on March 5, 2016, and the current one (version 1.0) was released in 2017. However, as of February 2019, the project is described as "not actively supported."

Rattle

Rattle is a popular user interface for data mining in the R programming language. It presents visual data summaries and statistical data and can model and transform data in supervised and unsupervised machine-learning models.

The beauty of Rattle is its feature that captures GUI interactions in scripts executable independently. It can be useful as a learning tool for developing R skills and later fine-tuning models in Rattle for more powerful data modeling capabilities.

It is a GUI for data mining with R. It presents statistical and visual summaries of data, transforms data so that it can be easily modeled, builds unsupervised and supervised machine-learning models from data, graphically presents model performance, and evaluates new data production deployment kits. A key feature is that all your interactions through the GUI are captured as an R script that can easily be run in R independently of the Rattle interface. Use it as a tool to learn and develop your skills in R and then build your initial models in Rattle to then tune them in R, which provides significantly more powerful capabilities.

Installing Rattle

There are two ways to install Rattle and RStudio. The first way is to install RStudio and then Rattle inside the RStudio platform. The second way is to install R and Rattle separately and then install RStudio later. We will first describe the procedures to do this using Rstudio. To do this without Rstudio, just type the commands directly into the R interface. For more details, see the Rattle book, DMRR.

Here are the operating system–specific instructions for getting started with Rattle. Ubuntu 22.04 is recommended, whether running with Parallels on macOS or WSL2 on Windows 11 or on VirtualBox. Find and download the correct RStudio installer version for your platform here. After downloading, run the installer and the installation process will start shortly.

For more information about RStudio, please visit "https://www.rstudio.com/products/rstudio/download/"

On Windows: Download, install, and run R from https://cran.r-project.org/bin/windows/base/old/4.1.3/R-4.1.3-win.exe

Installation Guide: Open RStudio and type the following command at the prompt (lower left corner of the screenshot below).

- install.packages("rattle"): Now enter the following two commands at the R command line. It loads the Rattle package into the main library and then starts Rattle.

- library (rattle): To start the Rattle GUI in RStudio, type the following command:

  - rattle(): If Rattle's standalone GUI appears, you're good to go!

It should be enough on Windows systems.

```
> install.packages("rattle")
> install.packages("https://access.togaware.com/
RGtk_2_2.20.36.2.zip", repos=NULL)
> library(rattle)
> rattle()
```

Installing on Ubuntu 22.04

Ubuntu is recommended by data scientist as a platform for data analysis. It can be installed on computers, new and old, as well as specific tablets and smartphones. It can replace the current operating system (Windows or OSX) or can be installed within VirtualBox on any platform, all for free. If installing within VirtualBox, enable VirtualBox guest addition and bidirectional cut/paste. It is also available in the Microsoft Windows Subsystem Store for Linux under Windows 10.

After installing Ubuntu on your computer,

```
$ sudo apt-get install wajig
```

Then perform the below commands:

```
$ sudo apt-get install r-recommended r-cran-XML
libgtk2.0-dev
$ R
> install.packages("rattle")
> install.packages("https://access.togaware.com/
RGtk2_2.20.36.2.tar.gz", repos=NULL)
> library(rattle)
> rattle()
```

Installing on Macintosh OS X (Leopard and Lion)

The definitive guide to installing Rattle on Mac as of June 2018 comes via Zhiya Zuo, where Yihui Xie noted that he has precompiled the RGtk2 and cairoDevice binaries so that we can easily install Rattle.

Eric Lin provided the following steps. To install dependencies at any point during the process, press yes. If you have previously failed to install, run "brew doctor" to eliminate problems. If there is a cleanup recommendation, run the brew cleanup. The steps are given below:

- Open up your Terminal.

- You will need to install XCode command line before installing R. We can install it by copying and pasting into terminal and then pressing enter: Xcode-select --install

- Go to home of brew website and copy and paste the command into the terminal from here: https://brew.sh/

- To install R, copy and paste this into your terminal and press enter: brew install r

- We need dependencies for Rattle: brew install gtk+

- Ensure Cairo doesn't exist yet (if you get an error that it doesn't exist that's fine): brew `uninstall --force Cairo`

- `--ignore-dependencies`

- Next: brew the cask install xquartz

- Next: brew install -- -with-x11 Cairo

- Now enter the command as capitalized: R

- Copy and paste these commands one at a time within R and when asked for CRAN press 4 (ignore any warnings and then note it may run for a while):

  ```
  install.packages("RGtk 2", dependencies = T)
  install.packages("cairoDevice", dependencies = T)
  install.packages("rattle", dependencies = T)
  ```

- If asked to install dependencies by source, make sure to select yes.

- It should install everything, including dependencies, successfully. If it type the following to close R: q()

- Enter back in the: R

- Load rattle library: library(rattle)

- Load rattle: rattle()

DESIGN OF THE R SYSTEM

The primary R system is available from Comprehensive R Archive Network, known as CRAN.[14] CRAN hosts many add-on packages that can be used to extend R's functionality. The "basic" R system you download from CRAN: Linux Windows Mac Source Code.

The R function is divided into several packages.

The "base" R system includes, among other things, the base package that is required to run R and contains the most basic functions. Other packages included in the "base" system include utils, stats, datasets, graphics, grDevices, grid, methods, tools, parallel, compiler, spline, tcltk, and stats4.

There are also "recommended" packages: boot, class, cluster, code tools, Foreign, KernSmooth, lattice, mgcv, nlme, rpart, survival, MASS, spatial, nnet, and Matrix. When you download a fresh installation of R from

CRAN, you get all of the above, which is the bulk of the functionality. However, there are many other packages available.

There are over 4000 packages on CRAN, developed by users and programmers around the world. There are also many packages associated with the Bioconductor project.

People often make packages available on their personal websites; there is no reliable way to keep track of how many packages are available this way.

There are a number of packages being developed on repositories like GitHub and BitBucket, but there is no reliable list of all these packages.

INTRODUCTION TO RSTUDIO

This section is an attempt to explain to beginners how to install, run, and use RStudio.[15] Contrary to many beginner assumptions, both R and RStudio are two different applications or software. First, you need to install R. RStudio is a standalone software that works with R to make R much more user-friendly with some useful features that make R programming easier and more efficient.

RStudio runs on Windows, Mac, and Linux and even over the web using RStudio Server. It is a good idea to learn how to use R, especially if you will be dealing with statistical operations where you can be expected to know how to use it.

How to Install RStudio
Installing R
Go to the R Project, website link is here "https://cran.r-project.org/," and download R for your operating system.

Installing RStudio
Go to RStudio, website link is here "https://www.rstudio.com/," and click on "Download RStudio" and follow the directions for your operating system.

DOWNLOAD AND INSTALL R

Precompiled binary distributions of the base system and added packages, Windows and Mac users most likely want one of these versions of R:

- Download R for Linux (Debian, Fedora/Redhat, Ubuntu)

- Download R for macOS

- Download R for Windows

R is included in many Linux distributions. In addition to the link above, you should consult the Linux package management system.

RStudio is an IDE for R. The IDE is a GUI where you can write your quotes, view the results, and also see the variables that are generated during programming.

- RStudio is available as both open-source and commercial software.

- RStudio is also available as desktop and server versions.

- RStudio is also available for various platforms such as Windows, Linux, and macOS.

INSTALLING RSTUDIO ON WINDOWS

To Install RStudio on Windows, we will follow the following steps.[16]

Step 1: First, you need to install and set up an R environment in your local machine. You can download the same from its official site, that is, "https://cran.r-project.org/" as per your system needs whether it is Windows, Linux, and macOS. You will get the link of each operating system from the above link.

Official site of R for download R environment.

You will get various .exe of different operating systems below:

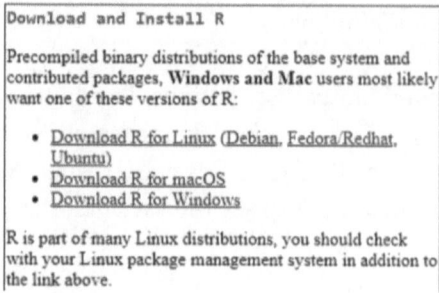

Different operating system .exe setup.

It will redirect you on this page (in Windows):

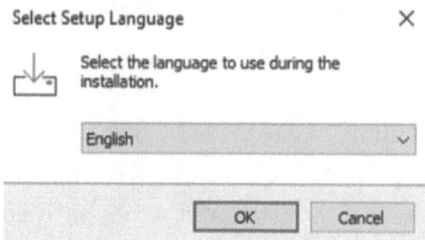

Redirect to the Windows version.

Click on any link to download the setup for your system. After downloading the .exe file. Double-click on it you will get pop up of "Select Setup Language" now select your language as per listed the figure is given below:

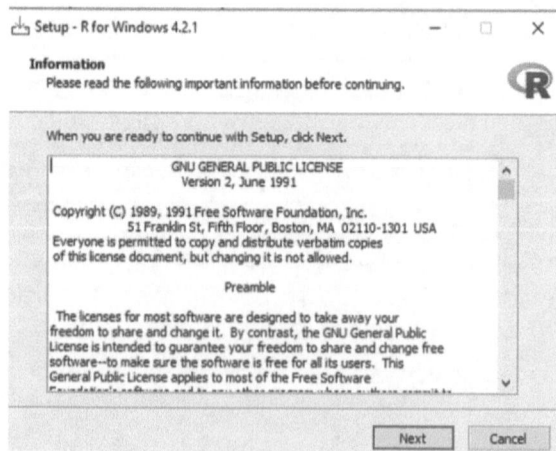

After download first window pop up.

Click on OK button and a new screen will appear agreeing the policy of the application; then click "Next" button.

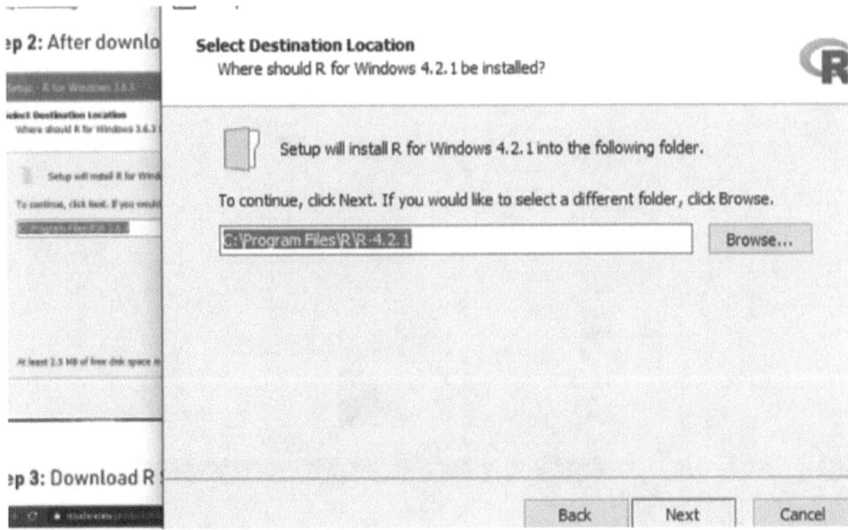

Agree policy application.

Now select your destination where you want to install and run R in your system the figure is given below:

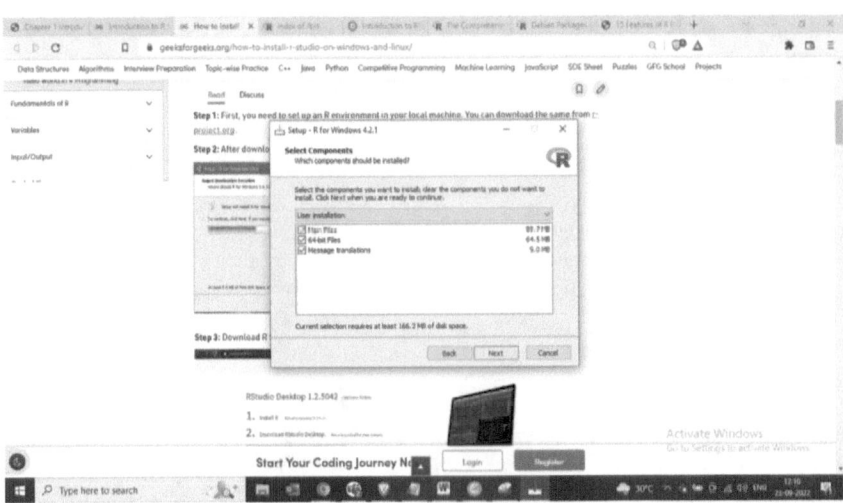

Destination of R folder.

After next, you can select the components you want in your R environment. The following figure is given below:

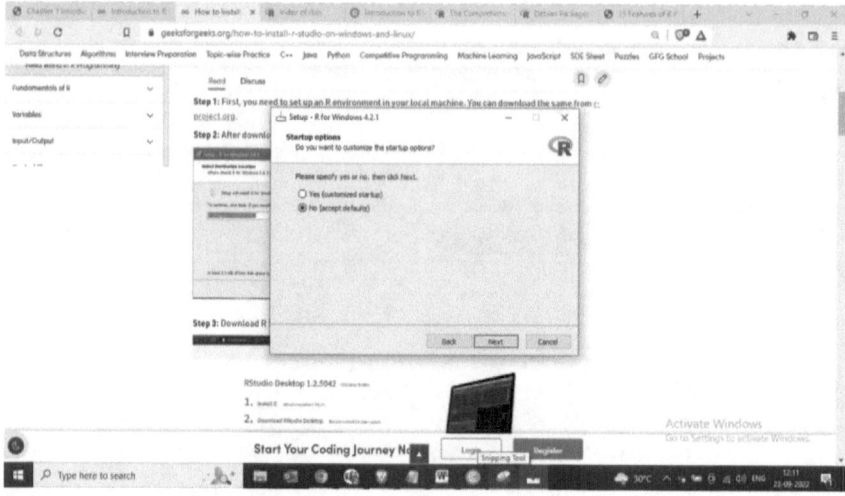

Select components of R.

Then you will get startup options in "YES" or "NO" which means whenever you will on your system this application will automatically turn on without opening my own the following figure is given below:

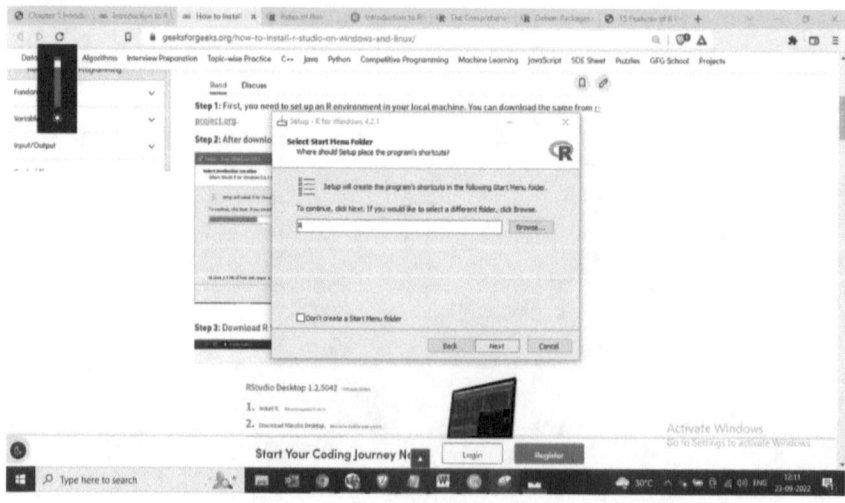

Startup options in R.

Select your start menu folder just by bowering the path of the particular folder as given below:

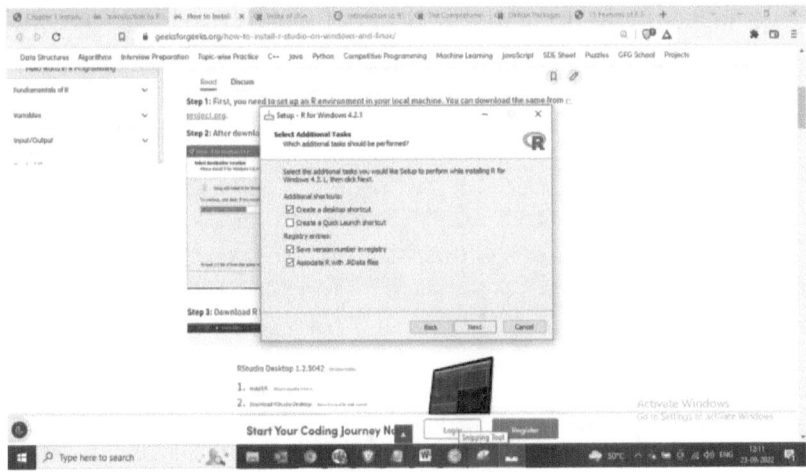

Select Start menu folder.

After that, select your R browsing folder means where you want to keep R files. You can click on the "Browse" button to choose the particular folder as such:

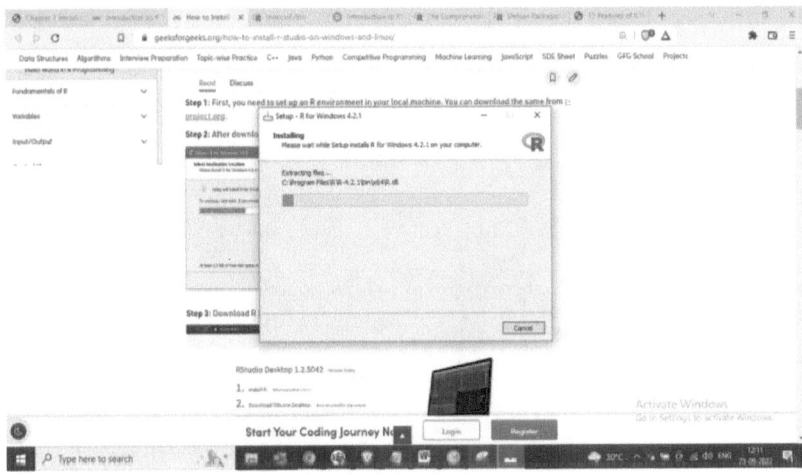

R browsing folder.

When you click "Next," the additional tasks window appears. Here you can add various information like you can add your application shortcut on the Desktop also add create a quick launch shortcut. On the other hand registry entries like save version number and associate R with .Rdata files such as

After all the steps, this will be your final step of installing the entire default file present. The software adds the other files in this step. Don't close your window which is given below:

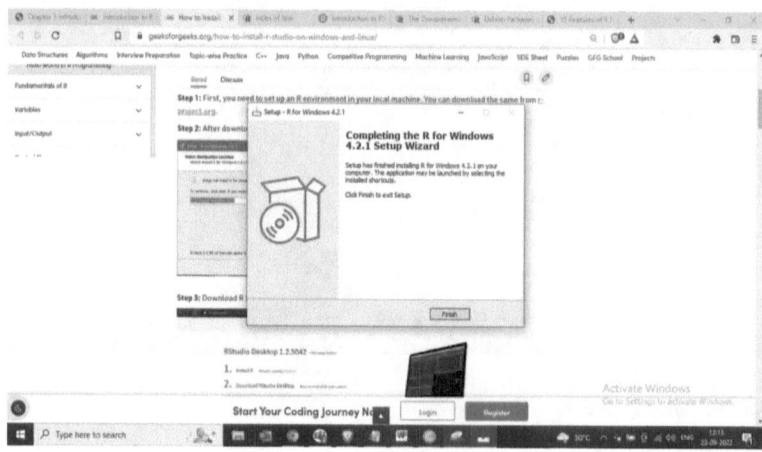

Final step of installation.

It may take a while once it gets done the application process will finish:

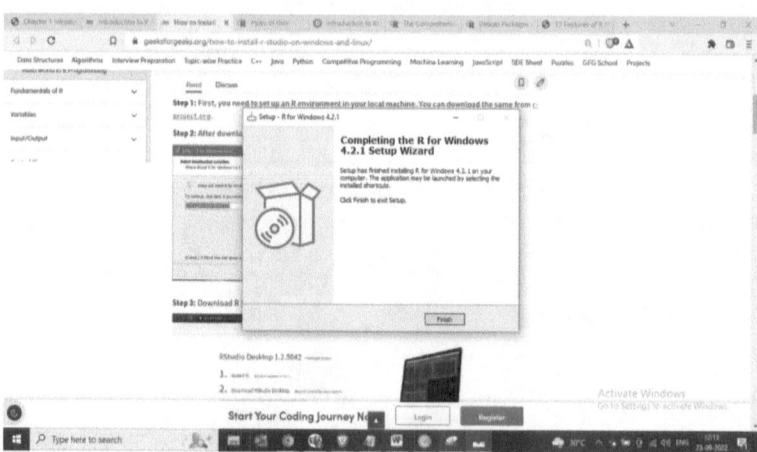

Last window of R setup installation.

Check if there is an "R" icon on the desktop of the computer. Then, double-click on the "R" icon to start R. If you cannot find an "R" icon.

Click on the "Start" button at the bottom left of a computer screen, and then choose then select Search Option, and start R by selecting "R" from the list.

On the other hand, after installation, you can find the program launch icon in the directory structure "C:\Program Files\R\R-3.4.1\bin\i386\Rgui.exe" under Windows Program Files. Clicking this icon will bring up the R-GUI, which is the R console for R programming.

R setup environment on the 32-bit version or the 64-bit version?

Most people don't need to worry about this. Obviously, the 64-bit version of R will not work on a 32-bit machine, but both the 32-bit and 64-bit versions of R run smoothly on 64-bit Windows.

RSTUDIO

RStudio is an open-source IDE that facilitates statistical modeling as well as graphical capabilities for R.[17] It uses the QT framework for its GUI functionality.

RStudio is an IDE for R. The IDE is a GUI where you can write your quotes, view the results, and also see the variables that are generated during programming.

- RStudio is available as both open-source and commercial software.

- RStudio is available as desktop and server versions.

- RStudio is available for various platforms such as Windows, Linux, and macOS.

RSTUDIO INSTALLATION

Now there are some steps to follow to get RStudio in your local system.

Step 1: Go to RStudio official site here is the link "https://www.rstudio.com/" as shown in the following figure you will get this:

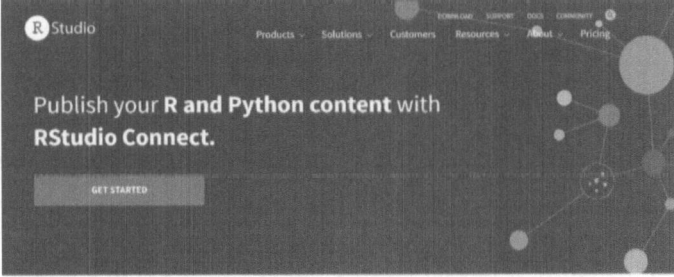

RS official site.

Step 2: Go upward, there you see Products > Open Source > RStudio. The following image will make you clear given below:

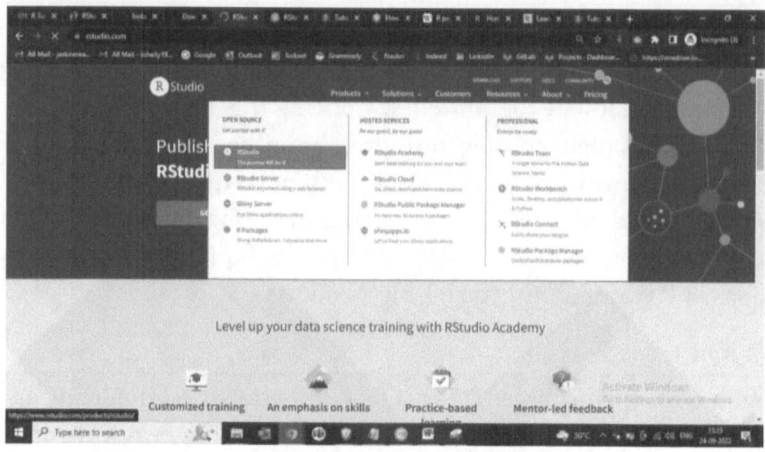

Here you get the downloading path of the RStudio.

Step 3: Then click on RStudio you will get this screen as given below:

There are two versions of RStudio:

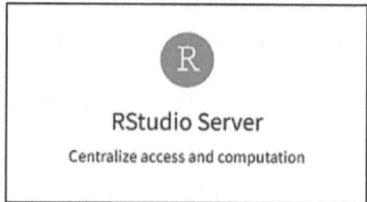

Two version of RStudio.

Step 4: Click on any version you want, but we recommend RStudio Desktop for beginners. You can purchase the Pro desktop version of it after working on the Open Source Editor.

RStudio Desktop

Features of the Open Source Edition:

- Access RStudio locally
- Syntax highlighting, code completion, and smart indentation
- Run R code directly from the source editor

- Quickly navigate to function definitions
- View content changes in real time with the Visual Editor
- It can Easily manage multiple working directories with projects
- Integrated R help and documentation
- Interactive debugger for diagnosing and fixing errors
- Extensive package development tools

Features of the RStudio Desktop Pro:

- Access RStudio locally
- Syntax highlighting, code completion, and smart indentation
- Run R code directly from the source editor
- Commercial licenses for organizations that cannot use AGPL software
- Access to priority support
- RStudio Professional Drivers
- Remotely connect directly to your instance of RStudio Workbench
- Quickly navigate to function definitions
- View content changes in real time with the Visual Editor
- It can easily manage multiple working directories with projects
- Integrated R help and documentation
- Interactive debugger for diagnosing and fixing errors
- Extensive package development tools

RStudio Server

Features are:

- Access via a web browser
- Move computation closer to the data
- Scale compute and RAM centrally

Features are:

- Access via a web browser
- Move computation closer to the data
- Scale compute and RAM centrally

All features are open-source Plus:

- Administrative tools
- Improved security and authentication
- Metrics and monitoring
- Advanced resource management
- Use RStudio, Python, Jupyter, and VSCode
- Native support for SAML and OpenID authentication for SSO

It also has RStudio Server and RStudio Workbench. The server version is free and workbench is paid.

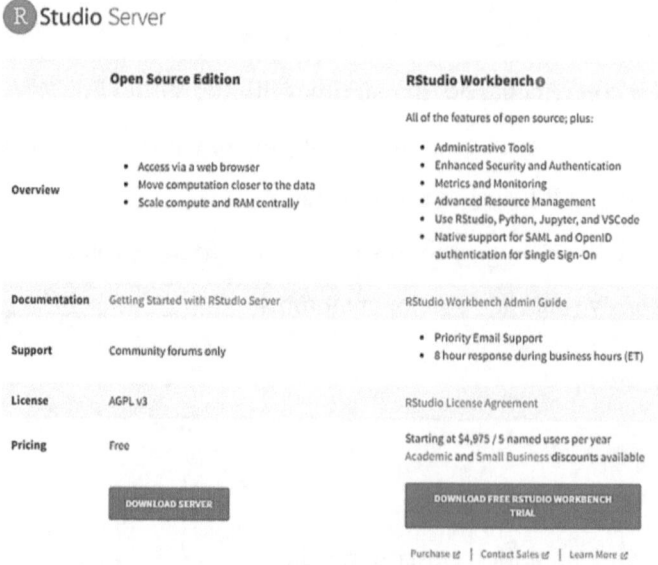

RStudio server and RStudio Workbench.

In short: There are two versions of RStudio – RStudio Desktop and RStudio Server. RStudio Desktop provides a facility to work on a local desktop environment, while RStudio Server provides access through a web browser.

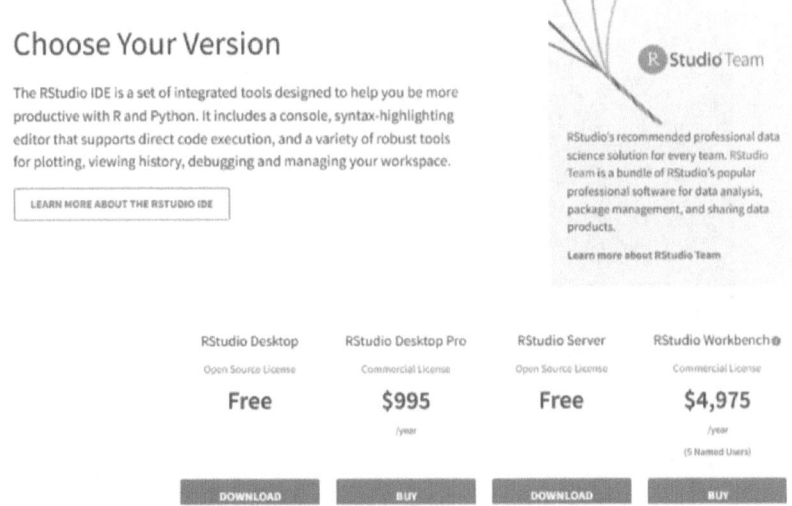

All version of RStudio.

Now Click on the RStudio Desktop version and download it.

Step 5: Next, find the file that was downloaded in your system and double-click it. It will be named something like RStudio-2022.07.2-576.exe. This will start the installation process.

RStudio-2022.07.2-576.exe

RStudio .exe file.

Step 6: Click next to continue when the install wizard opens the following image is given below:

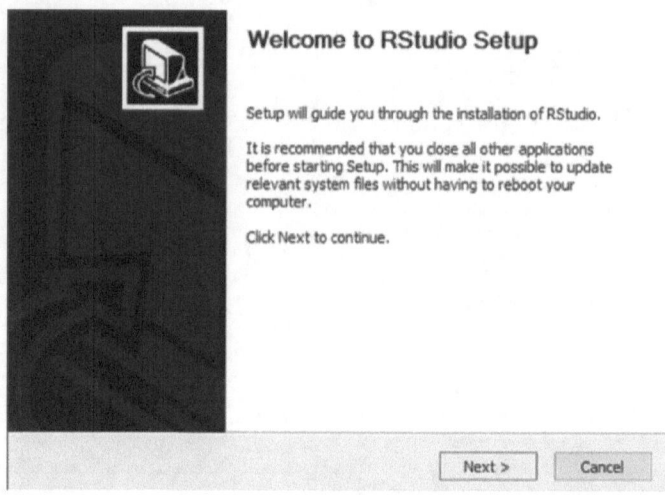

RStudio setup.

Step 7: Click next to accept the default install location. Also, you can change its location as per your space of drives as given below:

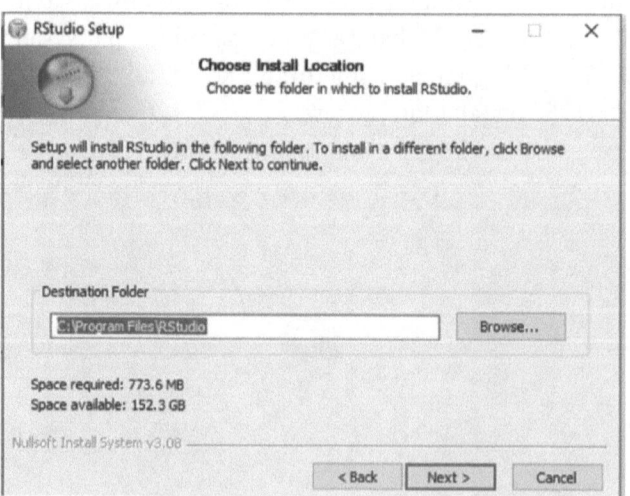

RStudio setup.

Step 8: Click Install to accept the default start menu folder and install RStudio.

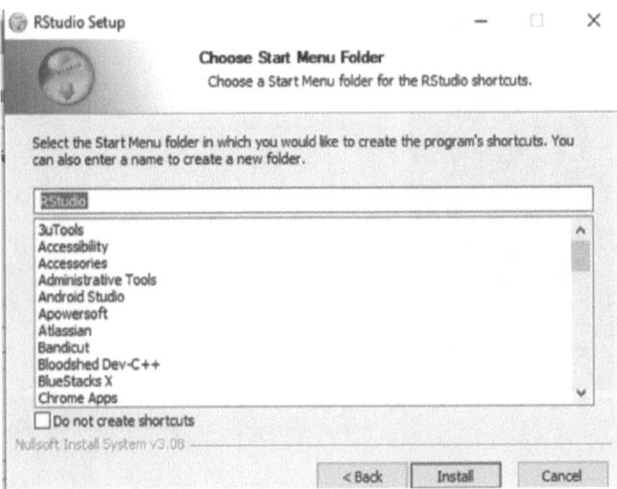

RStudio setup.

Step 9: Now it will be installed in your system but will take a while.

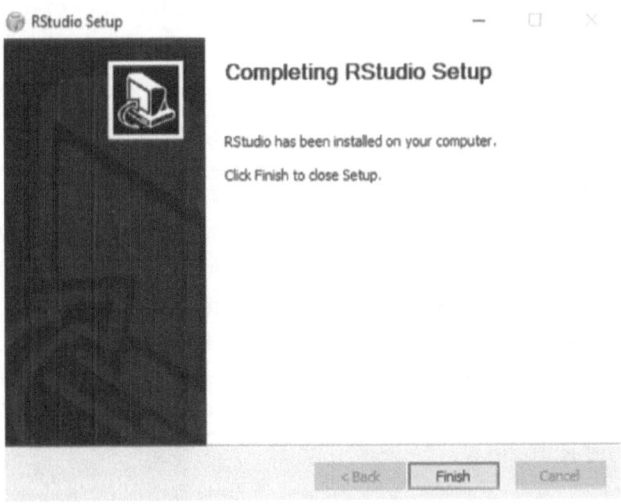

RStudio installation window.

Step 10: Next, click Finish to close the wizard.

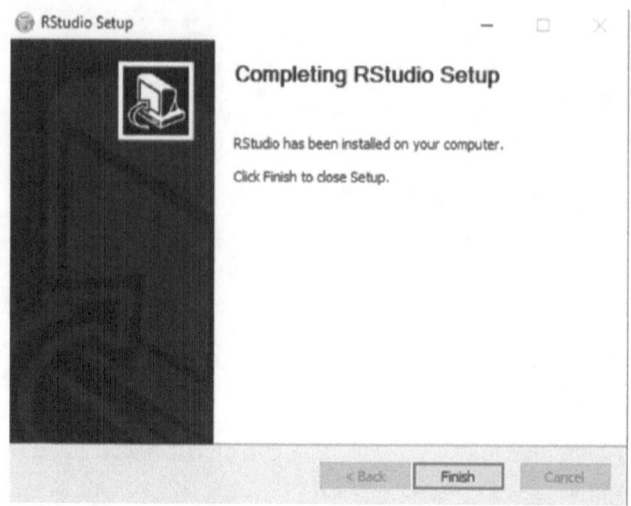

RStudio installation finish window.

Step 11: You can check if there is "RStudio" icon on your desktop. If it exists, double-click the "RStudio" icon to launch RStudio. If you can't find the "RStudio" icon.

Step 12: Click the "Start" button in the lower left corner of the computer screen and then select Search. Type RStudio there, and click RStudio to open it.

RSTUDIO USER INTERFACE

Initially, the RStudio window contains three smaller windows. For now, we'll focus on the large window on the left, the console window where R commands are entered. The next few sections show simple examples of using R. In these sections, we'll focus on small and uncomplicated data sets, but we'll be working with much more later in the book larger and more complex data sets. Read these sections on your computer running R and type R commands there to get comfortable with using the R console window and RStudio.

StatET for R

StatET is an Eclipse-based IDE for R programming.[18] It provides an unrivaled set of tools for writing R code and building packages. Features include an integrated R console, R object viewer and help, and its support for multiple local and remote installations. StatET is a plugin for the Eclipse IDE, so it can be combined with a number of other tools on the Eclipse platform. StatET is an open-source software that runs on most operating systems. It is because StatET is an Eclipse-based IDE, which makes this platform integrate a wide range of many other tools on the Eclipse platform.

Now no more hassle of writing codes because this IDE has a set of unrivaled tools for writing R code and building packages. In addition, it supports multiple local and remote installations.

Eclipse StatET™ is an Eclipse-based IDE for R.

1. It offers a set of advanced R coding and package-building tools. It includes a fully integrated R console

2. R script editors

3. It is an integrated R Graphics display

4. Its object browser is used to examine the objects that R has in memory

5. It is an integrated R help system

6. Its functionality to interact with multiple local and remote R installations

7. It is a visual debugger for R

8. It is editors and document processing support for Sweave/knitr and Rmarkdown documents

9. It is support for launching R CMD tools as an external launch configuration

10. and many more

Installation of StatET
For Windows

- Download the Eclipse Windows x86_64 installer available in the blue box titled "Try the Eclipse 2021-12 R Installer" you'll see at https://www.eclipse.org/downloads/packages/release/2021-12/r/

- Double-click the downloaded file.

- Finally, you will see a list of Eclipse IDE options; select "Eclipse IDE for Java Developers."

- Before selecting INSTALL, make sure you select JRE 17 next to Java 11 + VM.

- After the installation is complete, an Eclipse shortcut will appear on the desktop.

- Start Eclipse and install the Course Management Plugin installation as described below. If you don't install this plugin, you won't see the "Submit Project" button that allows you to submit projects from Eclipse.

- Eclipse does not come with an uninstaller. Some information regarding uninstalling Eclipse is provided at the end.

For Mac

- Download the Eclipse macOS x86_64 installer available in the blue box titled "Try the Eclipse 2021-12 R Installer" you'll see at https://www.eclipse.org/downloads/packages/release/2021-12/r/

- Double-click the downloaded file.

- Double-click the "Eclipse Installer" file that appears.

- Finally, you will see a list of Eclipse IDE options; select "Eclipse IDE for Java Developers."

- Before selecting INSTALL, make sure you select JRE 17 next to Java 11 + VM.

You won't find Eclipse in the Applications folder; use Spotlight to search for "Eclipse."

Start Eclipse and install the Course Management Plugin installation as described below. If you don't install this plugin, you won't see the "Submit Project" button that allows you to submit projects from Eclipse.

Eclipse does not come with an uninstaller. Deleting the eclipse folder in the /Users/USERNAME folder (where USERNAME corresponds to the username of the account) will remove eclipse. Deleting this Eclipse folder will not delete the workspaces. Multiple versions of eclipse can be located in the eclipse folder.

Features:

- Integrated R console is available

- R script editors

- R Graphics view

- Amazing support for running R CMD tools

- Object browser

- R help system

- Visual Debugger

KWard

R KWard is one of the most popular and widespread IDEs for developing projects from the R programming language. It is a highly scalable and easy-to-use IDE for R programming. R KWard was designed for R programmers. So, if you love writing R codes in R programming language, this IDE must be a great choice for you because it combines the power of R language along with statistical tools.

Function:

- It has a spreadsheet-like data editor

- Land review is available

- Managing R packages

- Workspace browser

- It has GUI dialogs for all kinds of statistics and graphs

- You can import data (e.g., SPSS, CSV)

- Compiling the code

- Code completion

- Syntax highlighting

- It can go through the history

Tinn-R

Tinn-R is a generic ASCII (and UNICODE) editor/word processor for the Windows operating system, very well integrated into R, with GUI and IDE characteristics. The purpose of the Tinn-R project is to facilitate the learning and use of all the capabilities of the R environment for statistical computing.

It has the features of a user interface and at the same time characteristics of an IDE. Its purpose is to facilitate learning R and provide an environment for statistical computing.

Features:

- The ability to communicate with R (Rgui and Rterm) and environment (send instructions, control, and receive interpretations)

- It has syntax highlight

- It can create and manage projects

- Bookmarks: lines and blocks

- It supports Latex

- It can work with files of unlimited length

- It can work on several documents at the same time. The multiple document interface (MDI) and/or tabbed document interface (TDI)

- It is single document window splitting and window splitting

- It supports macro (volatile)

- It can view file differences with color highlighting

- It supports UNICODE

- Templates to: Rscript, R doc, R HTML, R markdown, R noweb

- Portable (simple and Apps compatible)

R AnalyticFlow

R AnalyticFlow is a data analysis development environment for statistical computing. It also provides advanced features for R experts. These features allow data analysis processes to be shared between users with different levels of expertise. R AnalyticFlow works on Windows, Mac, and Linux and is free for any use.

It has a good user interface with advanced R features for R experts, which allows the sharing of analysis processes between multiple users of different R knowledge levels. It runs on Windows, Linux, and Mac operating systems. It is available under the GNU license.

R AnalyticFlow organizes data analysis processes in a workflow. The visualized processes can be easily and accurately reproduced using the mouse. Analytical workflows can be combined with related data and documents to create a project. With these features, R AnalyticFlow supports team sharing of analytical processes.

Why AnalyticFlow?

R AnalyticFlow supports Windows, Mac, and Linux, and the display language can be set to either English or Japanese. R AnalyticFlow also allows users in different environments to work collaboratively.

R AnalyticFlow is equipped with a wealth of support features. To name a few, the object browser to quickly confirm analysis results, object caching for saving and reusing processed results, debugging function, and automatic backup system. All these features will support your analysis strongly. It has well-equipped assist functions such as highlighting and code completion.

ADVANTAGES OF R PROGRAMMING

R success revolves around several advantages that it provides to both beginners and experts.[19] Here are some good advantages of R programming:

1. **Excellent for statistical calculations and analysis**

   R is a statistical programming language created by statisticians. So it excels in statistical calculations. R is the most widely used programming language for developing statistical tools.

2. **Open source**

   R is an open-source programming language. Anyone can work easily with R without any license or fee. Because of this, R has a huge community that contributes to its environment. We can contribute to the development by optimizing packages, developing new ones, and resolving issues. R is licensed under the General Public License, with copyright held by the R Foundation for Statistical Computing.

3. **A large selection of libraries**

   Massive support from the R community has resulted in a very large collection of libraries. R is known for its graphics libraries.

These libraries support and also enhance the R development environment. R has libraries with a large number of applications.

4. **Cross-platform support**

Cross-platform compatibility in R is machine independent. It supports cross-platform operation. So it is usable on many different operating systems. It enables programmers to develop software for several platforms by writing a program only once. R can run quite easily on Windows, can Linux, and Mac.

5. **It supports different data types**

R can perform operations on various types such as vectors, arrays, matrices, and various other data objects of various sizes. It is free and can be modified and customized according to user and project requirements. You can make improvements and add packages for additional features. R language is freely available. You can learn how to install it, download, and start practicing.

6. **Can perform data cleaning, data thrashing, and web scraping**

R can collect data from the net through web scraping and other means. It can also perform data cleaning. Data cleansing is the process of identifying and removing/correcting inaccurate or corrupted records. R is useful for data wrangling, which is the process of converting the raw data into a desired format for easier use.

7. **Powerful graphics**

R has extensive libraries that can create production-quality graphs and visualizations. These graphics can be static or dynamic in nature.

8. **Highly active community**

The R community is very active. There are users from all over the world to help and support you. Many of the latest ideas and technologies are emerging in the R community.

9. **Parallel and distributed computing**

By using libraries like DDR or multiDplyr, R can process large data sets using parallel or distributed computations.

10. **It does not need a compiler**

R is an interpreted language. It means that it does not need a compiler to turn the code into an executable program. Instead, it interprets the provided code into lower level calls and precompiled code.

11. **Compatible with other programming languages**

R is compatible with other languages such as C, C++, and FORTRAN. Other languages such as .NET, Java, and Python can also manipulate objects directly.

12. **It is used in machine learning**

R can also be useful for machine learning. Facebook does a lot of machine-learning research using R. Sentiment analysis and sentiment prediction are done using R. The best use when it comes to machine learning is when doing research or building one-off models.

13. **Can work with databases**

R includes several packages that allow it to interact with databases. Some of these packages such as Oracle, Open Database Connectivity Protocol), MySQL, and so on.

14. **Complex environment**

R has a very comprehensive development environment. It helps with statistical calculations and software development. R is an object-oriented programming language. It has a robust package called Rshiny that can build full-fledged web applications. It can also be useful for developing software packages.

15. **The array of packages**

R has a rich set of packages. It has over 10,000 packages in the CRAN repository which are now constantly growing. It provides packages for data science and machine-learning operations.

16. **Compatible with other programming languages**

While most of the functions are written in R, C, C++, or FORTRAN can be used for computationally intensive tasks. Java,. NET, Python, C, C++, and FORTRAN can also be used to manipulate objects directly.

17. **Data handling and storage**

R is integrated with all data storage formats, making data manipulation easy.

18. **Vector arithmetic**

Vectors are the most basic data structure in R, most other data structures are derived from vectors. R uses vectors and vector arithmetic and doesn't need many loops to process a large set of values. This makes R much more efficient.

DISADVANTAGES OF R PROGRAMMING

Let's get to the disadvantages of using R in the following:[20]

1. **Steep learning curve**

   R's syntax is very different from other languages, as are its data types. The learning curve for R is quite steep for a beginner. Although R is a bit difficult to get started with, data science enthusiasts still prefer to learn it because of R's amazing features. R is not that easy to use for a newbie. There are several easy-to-use GUIs for R that include point-and-click interactions, but they generally lack the gloss of commercial offerings.

2. **Poor-quality packages**

   CRAN contains over 10,000 libraries and packages. Some of them are also redundant. Due to the high quality, some packages may be of poor quality.

3. **Poor memory management**

   R commands are not related to memory management. R can take up all available space.

4. **Slow speed**

   Programs and functions are distributed in different packages. This makes it slower than alternatives like MATLAB and Python. The R language is much slower than other programming languages such as MATLAB and Python. Compared to other programming language, R packages are much slower.

   In R, algorithms are divided into different packages. Algorithms can be difficult to implement for programmers with no prior knowledge of packages.

5. **Poor security**

   R lacks basic security measures. So building web applications with it is not always safe.

6. **No dedicated support team**

   R does not have a dedicated support team to help the user with their problems and issues. But the community is quite large, so everyone helps each other.

7. **Flexible syntax**

R is a flexible language and there are no strict guidelines to follow. To avoid messy and complicated code, you need to follow proper coding standards.

CHAPTER SUMMARY

Above, you got all the basic knowledge of R like its history, Design of R Syntax, R Editors, Installation of R and Rstudio, and also the advantages and disadvantages. In the next chapter, we learn about data types in R.

NOTES

1. Data Science – https://ischoolonline.berkeley.edu/data-science/what-is-data-science-2/, accessed on September 21, 2022.
2. Data Science – https://www.datascience-pm.com/data-science-life-cycle/, accessed on September 21, 2022.
3. Introduction to R – https://www.r-project.org/about.html, accessed on September 21, 2022.
4. Origin of R – https://analyticsindiamag.com/introduction-to-basic-concepts-of-r-programming-language/, accessed on September 21, 2022.
5. Introduction R – https://static-bcrf.biochem.wisc.edu/courses/Tabular-data-analysis-with-R-and-Tidyverse/book/2-howRworks.html, accessed on September 21, 2022.
6. Usage – https://statanalytica.com/blog/uses-of-r/, accessed on September 21, 2022.
7. Uses R in Various Sectors – https://www.calltutors.com/blog/uses-of-r/, accessed on September 22, 2022.
8. Use of R in Research – https://techvidvan.com/tutorials/r-applications/, accessed on September 22, 2022.
9. Application of R in Real IT World – https://techvidvan.com/tutorials/r-applications/, accessed on September 22, 2022.
10. IT Sector Applications – https://data-flair.training/blogs/r-applications/#:~:text=R%20is%20one%20of%20the%20standard%20tools%20that%20is%20being,effective%20choice%20for%20these%20industries, accessed on September 22, 2022.
11. Reason to Use R – https://www.geeksforgeeks.org/r-programming-language-introduction/, accessed on September 22, 2022.
12. R Package – https://en.wikipedia.org/wiki/R_package#:~:text=The%20Comprehensive%20R%20Archive%20Network%20(CRAN)%20is%20R's%20central%20software,binaries%20for%20Windows%20and%20macOS, accessed on September 22, 2022.
13. IDE and Coding Editors of R – https://www.dunebook.com/best-r-programming-ide/, accessed on September 22, 2022.
14. Design of R – https://bookdown.org/rdpeng/rprogdatascience/history-and-overview-of-r.html, accessed on September 24, 2022.

15. Introduction to R – https://datascienceplus.com/introduction-to-rstudio/, accessed on September 23, 2022.

16. R installation steps – https://makemeanalyst.com/r-programming/installing-r-on-windows/, accessed on September 24, 2022.

17. Rstudio – https://data-flair.training/blogs/rstudio-tutorial/, accessed on September 24, 2022.

18. Text Editors – https://mytrashcode.com/best-ide-r, accessed on September 23, 2022.

19. Advantages and Disadvantages – https://www.javatpoint.com/r-advantages-and-disadvantages, accessed on September 21, 2022.

20. Disadvantages – https://www.solutions4statistics.com/post/2017/07/03/advantages-and-disadvantages-of-using-r-programming, accessed on September 21, 2022.

## CHAPTER 2

Handling Data with R

IN THIS CHAPTER

> ➤ RStudio Application User Interface

> ➤ Basic syntax in R programming

> ➤ Comments

> ➤ Keywords

> ➤ Keyboard Shortcut

In the last chapter, we did a basic introduction on R with its history, version, advantage, and disadvantages. Now we have the main concept of R is data and what type of data we can use in R programming language. Before starting, we have an overview of RStudio that how the application looks like. Then, the data of R will be covered.

GETTING STARTED WITH THE RSTUDIO APPLICATION

RStudio is an open-source tool for programming in R. It is a flexible tool that helps you to create readable analyses and keeps your code, images, comments, and charts together in one place. It's worth knowing about RStudio's capabilities for data analysis and programming in R. It is a program that runs R and provides extra tools that are helpful while writing Rits code, kind of how an operating system can run a web browser.

DOI: 10.1201/9781003358480-2

Using RStudio Software for data analysis and programming in R provides many advantages. Here are some examples of what RStudio provides:

- An intuitive interface that allows us to monitor stored objects, scripts, and images

- A text editor with features such as color-coded syntax to help us write clean scripts

- The autocomplete feature saves time

- Tools for creating documents containing project code, notes, and visuals

- It is a dedicated project folder to keep everything in one place

When you open RStudio for the first time, you will probably see a layout like this:

RStudio user interface.

If you open RStudio for the first time, the background color will be white.[1] You can change it. When we open RStudio, R also starts. A common error of new users is to open R instead of RStudio. To open RStudio, locate RStudio on the desktop and pin the RStudio icon to your preferred location (e.g., desktop or toolbar).

When you open RStudio, you will notice many different windows, each with some tabs. It can be overwhelming at first, but you need to use a few to be productive, the ones you want similar to other system programs you probably already use.

THE CONSOLE

Let's start by introducing some of the console's features. Console is a tab in RStudio application where we can run R code.

Note that the window pane where the console is located has three tabs: console, terminal, and jobs (it may vary depending on the version of RStudio you are using). Now let's focus on the console.

When we open RStudio, the console can contain information about the version of R we are working with the application. Scroll down and try entering a few expressions like this. Press Enter to display the result.

Example:

```
R version 4.2.1 (2022-06-23 ucrt) -- "Funny-
Looking Kid"
Copyright (C) 2022 : The R Foundation for
Statistical Computing
Platform: x86_64_w64-mingw32/x64 (64-bit)

R is free software and comes with the ABSOLUTELY
NO WARRANTY.
You are welcome to redistribute it under certain
conditions.
Type 'license()' or 'licence()' for distribution
of the contributors.
Type 'contributors()' for the more information and
'citation()' on how to cite R or R packages in
publications.

Type 'demo()' for the some demos, 'help()' for the
on-line help, or
'help.start()' for the HTML browser interface to
help.
Type 'q()' to quit R.

> a = 20
> b = 10
> result = a + b
> result
[1] 30
> c <- a
> c
[1] 20
```

```
Console   Terminal ×   Background Jobs ×                                    ⊟ ⏏
R   R 4.2.1 · ~/ ⌀

R version 4.2.1 (2022-06-23 ucrt) -- "Funny-Looking Kid"
Copyright (C) 2022 The R Foundation for Statistical Computing
Platform: x86_64-w64-mingw32/x64 (64-bit)

R is free software and comes with ABSOLUTELY NO WARRANTY.
You are welcome to redistribute it under certain conditions.
Type 'license()' or 'licence()' for distribution details.

R is a collaborative project with many contributors.
Type 'contributors()' for more information and
'citation()' on how to cite R or R packages in publications.

Type 'demo()' for some demos, 'help()' for on-line help, or
'help.start()' for an HTML browser interface to help.
Type 'q()' to quit R.

> |
```

Console tab in RStudio.

As we can see, we use the console to test the code immediately. When we type an expression like 7 + 5, we will see the output below after pressing enter.

```
> 7 + 5
[1] 12
```

We can save the output of this command as a variable. Here we have named our variable result as:

```
Final_result <- 1 + 2
```

<- is called the assignment operator. This operator assigns values to variables. The above command is translated into a sentence as

The result variable gets the value one plus two.

RStudio has the keyboard shortcut for typing the assignment operator <-:[2]

- For Mac OS X: Option + -

- For Windows/Linux: Alt + -

The R console is where you enter R commands and is the bottom left window in RStudio. This is the same way you interact with R on the command line or terminal. In other words, the "Console" tab in the left window is the part of RStudio that is actually R itself; all others are optional tools.

The Text Editor

The top left window is a plain text editor like Notepad or Text Edit. The "Plain text" means no fonts or formatting, unlike a program such as Microsoft Word. You can have various files open at once and they will appear in tabs. It's depending on the type of the file being edited (i.e. file extension), there will be different tools and behavior, but it's just plain text.

Console tab options.

File Browser Tab

The default tab in the lower right corner of the window is the basic file browser. Here you can open, delete, and rename files. It's not as well developed as your operating system's file browser, and it's there most of the time so you don't have to switch apps to manage your files. You can ignore the remaining tabs (Plots, Packages, Help, and Viewer) for now as they usually open automatically when relevant.

File Browser tab options.

"Environment" Tab

The "Environment" tab is on the upper right window which lists the variables and functions present in the current R session. However, it does not include the function/data in the loaded packages (unless you select a package from the drop-down menu called "Global Environment").

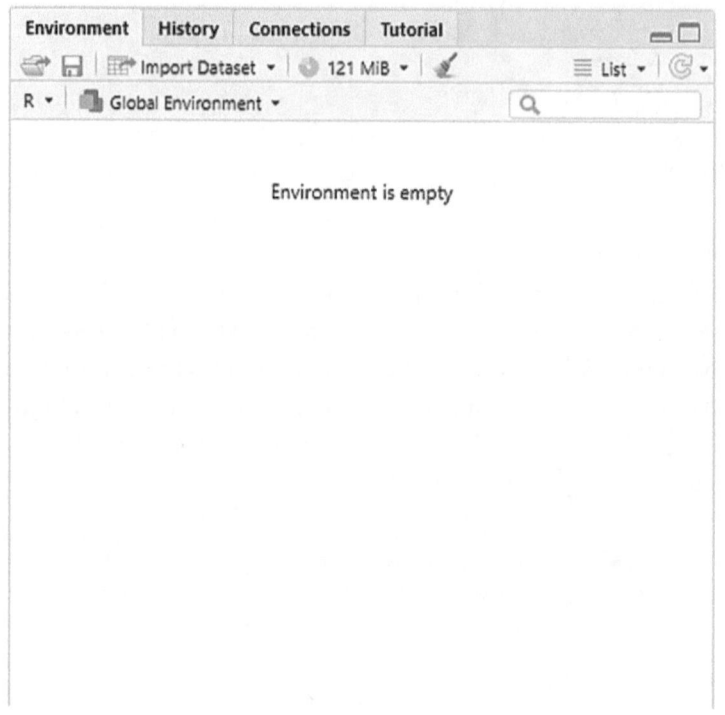

Environment tab options.

We will see all the objects we created, for example, the result, under the values in the Environment tab. We notice that the value 3 stored in the variable is displayed.

Too many named objects in a global environment sometimes cause confusion. We might want to remove all or some items. To remove all objects on the screen, click the "broom" icon at the top right corner of the window.

While programming session in R, any variables we define or data we import and store in a data frame are stored in our global environment. In RStudio application, we can see the objects in our global environment on the Environment tab in the upper right part of the user interface.

Code Completion

RStudio IDE supports automatic code completion using the Tab key. For example, if you have any object called pollResults in your workspace, you can enter a query and then Tab and RStudio will automatically complete the entire object name. Code completion also provides inline help for functions whenever possible. For example, if you type hi and then press Tab, you'll see.

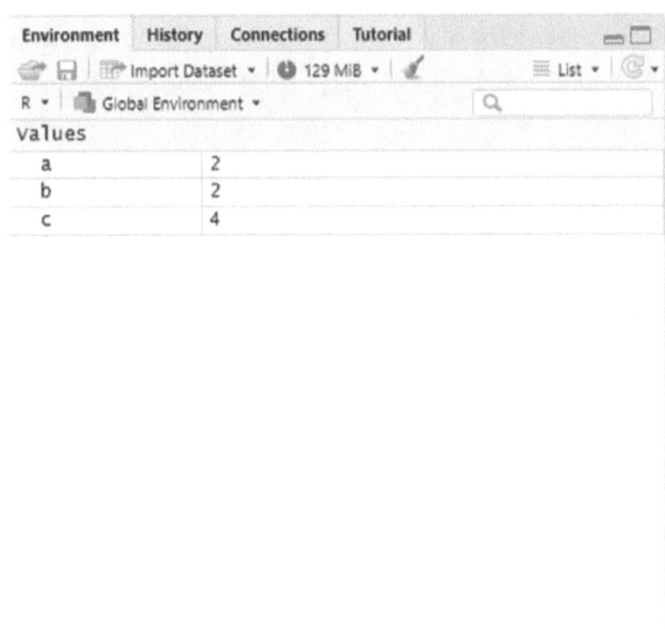

Tab option RStudio (list of hi function).

Code completion also works for function arguments, so if you typed subset or any other function and then pressed Tab, you'd see the following:

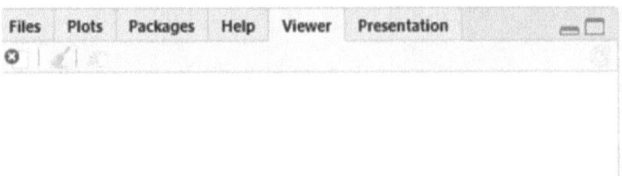

Tab option RStudio (list of hi function).

Retrieving Previous Commands

When working with R, you'll often want to re-execute a command you previously entered.[3] As with the standard R console, the RStudio console supports the ability to invoke previous commands using the arrow keys:

- Up – It brings up the previous command(s)

- Down – The opposite of Up

If you need to see a list of your recent commands and then select a command from that list, you can use Ctrl+Up to display the list.

Console Title Bar

This screenshot illustrates a few more features that the console header provides

- Display the current working directory.

- Ability to interrupt R during a long calculation.

- Minimize and maximize the console in relation to the Source panel (using the top right buttons or double-clicking the header).

Keyboard Shortcuts

In addition to the history and code completion-oriented hotkeys described above, a wide variety of other hotkeys are available. Some of the more useful abbreviations include:

- Ctrl+1 – It moves focus to resource editor

- Ctrl+2 – It moves focus to console

- Ctrl+L – It clear the console

- Esc – It is for Interrupt R program and application

Here is the complete list of keyboard shortcuts in the R programming:[4]

Basic Syntax in R Programming

R is the most famous language used for statistical computing and data analysis, supported by more than 10,000+ free packages in the CRAN repository. Like other programming languages, R has a specific syntax that is important to understand if you want to take advantage of its powerful features.

In this section, R is already installed on your system. We will be using RStudio but can also use the R command line by entering the following command at the command line.

```
$ R
```

We can see that "Hello, world!" is printed on the console. Now, we can do the same using the print() function which prints the console. We usually write our code in scripts, in which R is called scripts. To create it, write the code below in a file and save it as my_file.R, then run it in the console by typing:

```
Rscript my_file.R
```

R Syntax

A program in R consists of three things: variables, comments, and keywords. Variables are used to store data, comments are used to improve code readability, and keywords are reserved words that have a specific meaning to the compiler.

Variables in R

We used to write all our code in print(), but we do not have a way to resolve them to perform other operations. This problem can be solved by using variables; like any other programming language, these are named dedicated memory locations that can keep any type of data. In R, assignments can be marked in three ways:

1. = (Simple assignment)

2. <- (Left Assignment)

3. -> (Right assignment)

Creating Variables in R

Variables are containers for storing data values.[5] R doesn't have a command for declaring a variable. A variable is created the moment a value is assigned to it. To assign a value to any of a variable, use the <- sign.

To output (or print) the value, just type the variable name. The example is given below.

Print/Output Variables

Compared to many other programming languages, you don't need to use the function to print/output variables in R. Just specify the variable name:

Comments

Comments are a way to improve code readability and are intended only for the user, so they are ignored by the interpreter.[6] Only single-line comments are available in R, but we can use multiline comments using a simple trick shown below. Single-line comments can be written using # at the beginning of the command. More precisely, information that a programmer should be interested in has nothing to do with the logic of the code. They are ignored by the compiler and therefore never reflected in the input.

Here, we have an important question that how does the translator know whether a given statement is a comment?

The answer is that all languages use a symbol to indicate a comment, and this symbol, when encountered by the compiler, helps it distinguish between a comment and a statement.

Comments are used for the following purposes:

- The code can readable

- An explanation of the project's code or metadata

- Prevent code from running

- Include resources

- It tries to give an idea of what your block of code is doing. So that your teammates working on a specific project can immediately understand just by reading the comments.

Types of Comments

There are three types of comments supported by languages, namely

- Single-line comments: These are comments that only need one line.

- Multiline comments: These are comments that require more than one line.

- Documentation comments: These are comments that are usually designed to quickly locate the documentation.

Note: R does not support multiline comments and documentation comments. It only supports single-line comments with the "#" symbol.

Comments in R

As mentioned in the note above, R does not currently support multiline comments and documentation comments. R provides its users with single-line comments to add information about the code.

Single-Line Comments in R

Single-line comments are comments that only require one line. They are usually designed to explain what a single line of code does and what it is supposed to produce to help someone referencing the source code.

Like Python, single-line comments, any command beginning with "#" is a comment in R.

Syntax:

```
# comment statement
# R program to add two numbers
```

Here we assign values to variables a with value 9 and b with 4

```
> a <- 10
> b <- 14
```

Printing sum

```
> print(a+b)
[1] 24
```

As mentioned earlier, R does not support multiline comments but to make the commenting process easier, R allows you to comment multiple lines at once. There are two ways to add multiple single-line comments in R Studio, such as

1. Use the cursor to select multiple lines to comment, then use the key combination "control + shift + C" to comment the selected lines.

2. The second way is to use the graphical user interface; use the cursor to select the lines you want to comment and click "Code" in the menu; a pop-up window will appear in which we need to select "Comment/ Uncomment." Lines will appropriately comment or uncomment the lines you have selected.

Keywords in R

Keywords are words reserved by the program because they have a special meaning, so a keyword cannot be used as a variable name, function name, etc. Like in C, C++, and Java, there is a set of keywords in R. A keyword cannot be used as a variable name. The keywords are also called as "reserved names." There are the following keywords as per reserved or help(reserved) command:[7]

In this section, you get to know about various keywords used in R programming with the help of examples.

1. **if:** If statement consists of the Boolean expression and followed by one or more statements. In R, the if statement is the simplest conditional statement used whether a statement block will be executed or not. Decision-making is the most important part of programming. It can be achieved in programming using the if...else conditional statement.

   The basic syntax of if statement is:

   ```
   if (expression) {
   # Write you code here
   }
   ```

   If test_expression is TRUE, the statement is executed. But if it is FALSE, nothing happens. Test_expression here can be a Boolean or a numeric vector, but only the first element is considered. In the case of a numeric vector, 0 is taken as FALSE, the rest as TRUE.

   Example:

   ```
   > a <- 15
   > if (a < 18)
   + print(" Age is less than 18 ")
   [1] " Age is less than 18 "
   ```

   Another example using else with if given below:

   ```
   # It is program to check the two number whether
   # it is greater or smaller
   ```

```
# and print some message using print().

a <- 25
b <- 20

# Set the is-else statement
if (a > b)
+ print (' A is greater than B ')
+ } else
+ {
 + print (' B is greater than A')
+ }
```

The output of code is like

```
[1] " A is greater than B"
```

2. if–else

Example:

```
# It is program to check the two number whether
# it is greater or smaller
# and print some message using print().

a <- 25
b <- 20

# Set the is-else statement
if (a > b)
+ {
+ print (' A is greater than B ')
+ } else
+ {
 + print (' B is greater than A')
+ }
```

The output will be like this:

```
[1] " A is greater than B"
```

3. **repeat:** The repeat keyword is used to repeat a block of code repeatedly. There is a repeat loop in R, and there is no condition to exit the loop in this loop statement. To end the cycle, we use the break command. We have to put the condition explicitly in the loop body and

use the break statement to end the loop. Failure of the loop will result in an infinite loop.

The syntax of repeat loop is given below:

```
repeat {
statement
}
```

Example:

```
> x <- 1
> repeat {
+ print (x)
+ x = x + 1
+ if ( x == 6){
+ break
+ }
+ }
```

The output of the code is given:

```
[1] 1
[1] 2
[1] 3
[1] 4
[1] 5
```

In R console, you can write code above like this:

```
> x <- 1
> repeat {
+       print (x)
+       x = x + 1
+       if ( x == 6){
+           break
+       }
+ }
[1] 1
[1] 2
[1] 3
[1] 4
[1] 5
>
> |
```

repeat keyword.

4. **else**

```
> if (a == b)
+ {
+ print(" The value of a and b is same")
+ } else
+ {
+ print(" The value is different ")
+ }
[1] " The value of a and b is same "
```

5. **while:** It is a type of control statement that will execute a statement or set of statements repeatedly unless a given condition becomes false. It is also an input-controlled loop; in this loop, the test condition is first tested, then the loop body is executed; the loop body would not be executed if the test condition is false.

The syntax of while loop is given below:

```
while ( condition )
{
    statement
}
```

Example:

```
> a <- 5
> while (a <= 10)
+ {
+ print(a)
+ a = a + 1
+ }
[1] 5
[1] 6
[1] 7
[1] 8
[1] 9
[1] 10
```

6. **Function:** A set of commands that are organized together to perform a specific task is called a function.[8] R provides a number of built-in functions and allows the user to create custom functions. Functions are used to perform tasks in a modular approach.

Functions are used to avoid repeating the same task and reduce complexity. In order to understand and maintain our code, we logically divide it into smaller parts using a function. The function should be

- Written to accomplish a given task.

- It may or may not have arguments.

- They contain the body in which our code is written.

- It may or may not return one or more output values.

Here is the syntax of function in R programming:

```
new.func_name <- function(arg_1, arg_2, ...) {
    # here you can write function body
}
```

Example:

```
> new.function <- function(){
+ for(i in 1:10){
+ print(i+2)
+ }
+ }
> new.function()
```

Here is the output of the code:

```
[1]  3
[1]  4
[1]  5
[1]  6
[1]  7
[1]  8
[1]  9
[1]  10
[1]  11
[1]  12
```

There are various properties used in the R function such as:

- Function name: The function name is the actual name of the function. In R, a function is stored as an object with its own name.

- Arguments: In R, the argument is a placeholder. In a function, arguments are optional, which means that the function may or may not contain arguments and those arguments may also have default values. When we call the function, we pass a value to the argument.

- The body of the function: The body of a function contains a set of statements that define what the function does.

- The return value: It is the last expression in the function body to be evaluated.

7. **For:** The For loop in the R programming language is useful for iterating over the elements of a list, data frame, vector, matrix, or any other object. This means that a for loop can be used to repeatedly execute a group of statements depending on the number of elements in the object. This is an input-controlled loop; in this loop, the test condition is first tested, then the loop body is executed; the loop body would not be executed if the condition is false.

   For loop syntax:

```
for (var in vector) {
    # coding statement(s)
}
```

Example:

```
> for (i in 1:10)
+ {}
> for (i in 1:10)
+ {
+ print(i+2)
+ }
[1]  3
[1]  4
[1]  5
[1]  6
[1]  7
[1]  8
[1]  9
[1]  10
[1]  11
[1]  12
```

8. **Next:** The next statement in R is used to skip all the statements in the loop and continue executing the program. On the other hand, it is a statement that skips the current iteration without terminating the loop.

"next" is a loop control command just like the break command. But the "next" statement does the opposite of the break statement, instead of terminating the loop; it forces the next iteration of the loop.

Example:

```
> Val <- 6:15
> for(i in val)
+ {
+ if(i == 10)
+ {
+ next
+ }
+ print(i)
+ }
[1] 6
[1] 7
[1] 8
[1] 9
[1] 11
[1] 12
[1] 13
[1] 14
[1] 15
```

9. **Break:** In R, the break statement is used to interrupt execution and exit a loop immediately.[9] In nested loops, interrupt the outputs from the innermost loop only and direct the transfer to the outer loop.

It is useful to control and control the flow of program execution. We can use it for different loops like: for, repeat, and so on.

There are basically two uses of the break command, which are as follows:

• When a break statement is inside a loop, the loop is immediately terminated and program control is resumed at the next statement after the loop.

• It is also used to terminate a case in a switch statement.

Example:

```
> var <- 6:15
> for(i in val)
+ {
+ if (i==13)
+ {
+ break
+ }
+ print(i)
+ }
[1] 6
[1] 7
[1] 8
[1] 9
[1] 10
[1] 11
[1] 12
```

10. **TRUE:** TRUE/FALSE: The keywords TRUE and FALSE are used to express the Boolean values true and Boolean false. If the given statement is true, the interpreter returns true; otherwise, the interpreter returns false.

Example:

```
# A simple R program
# to find the TRUE/FALSE

# Sample values
> x <- 14
> y <- 15

# Comparing two values
> z = x > y
> p = x < y
> a = x = y
> b = x == y
```

The output of the code is given below:

```
# print the logical value
> print(z)
```

```
[1] FALSE
> print(a)
[1] 15
> print(a)
[1] 15
> print(b)
[1] TRUE
```

11. **FALSE**

Example:

```
# A simple R program
# to find the FALSE

# Sample values
> x <- 12
> y <-20

# Comparing two values
> z = x > y
> a = y < x
> b = x == y
```

The output of the code is given below:

```
# print the logical value
> print(z)
[1] FALSE
> print(a)
[1] FALSE
> print(b)
[1] FALSE
```

12. NULL: NULL represents a NULL object in R.[10] NULL is used primarily to represent zero-length lists and is typically returned by expressions and functions whose value is undefined. To define a NULL value, assign it to a variable and that variable will become NULL.

```
RV <- NULL
RV
typeof(RV)
```

In R, NULL (uppercase) is a reserved word and can also result from importing data with an unknown data type.

R provides two functions to deal with NULL.

- is.null()

- as.null()

13. **NA:** It is a logical constant of length 1 and is a missing-value indicator. NA (uppercase) is a reserved word and can be converted to any other data type vector (except raw) and can be a product when importing data. NA and "NA" (as presented as a string) are not interchangeable. NA stands for not available.

14. **NaN:** It stands for "Not A" number and is a logical vector of length 1 and applies to numeric values as well as the real and imaginary parts of values, but not integer vector values. NaN is a reserved word.

15. **Inf:** Inf stands for infinity (or negative infinity) and is the result of storing either a large number or a product that results from dividing by zero. Inf is a reserved word and in most cases—a product of R calculations and therefore very rarely a product of data import. Its infinite tells that the value is not missing and the number.

    All four null and missing data types have Boolean functions available in base R; returns TRUE/FALSE for each specific function: is.null(), is.na(), is.nan(), and is.infinite().

Example:

```
# Here we perform code of NULL, NA, Inf, NaN.
> x <- 12
> y <-20
> z = x > y
> a = y < x
> b = x == y
> print(z)
[1] FALSE
> print(a)
[1] FALSE
> print(b)
[1] FALSE
> #populating variables
> a <- "NA"
> b <- "NULL"
> c <- NULL
> d <- NA
```

```
> e <- NaN
> f <- Inf
> ### Check if variables are same?
> identical(a,d)
[1] FALSE
> # [1] FALSE
> # NA and NaN are not identical
> identical(d,e)
[1] FALSE
> # [1] FALSE
> ###checking length of data types
> length(c)
[1] 0
> # [1] 0
> length(d)
[1] 1
> # [1] 1
> length(e)
[1] 1
> # [1] 1
> length(f)
[1] 1
> # [1] 1
> ###checking data types
> str(c); class(c);
 NULL
[1] "NULL"
> #NULL
> #[1] "NULL"
> str(d); class(d);
 logi NA
[1] "logical"
> #logi NA
> #[1] "logical"
> str(e); class(e);
 num NaN
[1] "numeric"
> #num NaN
> #[1] "numeric"
> str(f); class(f);
 num Inf
[1] "numeric"
> #num Inf
> #[1] "numeric"
```

NA values represent the missing values in R.[11] They are awesome because they are natively baked right into R. There are actually many different flavors of NA values in R:

- NA is logical (NA)

- NA_integer_ are characters (NA_integer_)

- NA_character_ are integer values (NA_character_)

- NA_real_ is double (values with decimal points) (NA_real_)

- NA_complex_ for complex values (like 1i) (NA_complex_)

Here is an example of NA in different properties:

```
> is.na(NA)
[1] TRUE
>
> #> [1] TRUE
>
> is.na(NA_character_)
[1] TRUE
>
> #> [1] TRUE
>
> is.character(NA_character_)
[1] TRUE
>
> #> [1] TRUE
>
> is.double(NA_character_)
[1] FALSE
>
> #> [1] FALSE
>
> is.integer(NA_integer_)
[1] TRUE
>
> #> [1] TRUE
>
> is.logical(NA)
[1] TRUE
>
> #> [1] TRUE
```

This means that these NA values have different properties, even if they print as NA when printed; they are character, complex, or whatever.

Among these words, if, else, repeat, while, function, for, in, next, and break are used for conditions, loops, and user-defined functions.

- They form the basic building blocks of programming in R.

- TRUE and FALSE are Boolean constants in R.

- NULL represents the absence of a value or an undefined value.

- Inf is for "infinity," such as when 1 is divided by 0, while NaN is for "Not a number," such as when 0 is divided by 0.

- NA stands for "not available" and is used to represent missing values.

- R is a case-sensitive language, which means that TRUE and True are not the same thing.

- While the former is a reserved word denoting a Boolean constant in R, the latter can be used as a variable name.

Knowing the RStudio keyboard shortcuts will save a lot of programming time.[12] RStudio provides dozens of useful shortcuts that can access through the menu at the top: Tools > Hotkey Help. Another way to access RStudio keyboard shortcuts is the Shortcut! To access shortcuts, type Option + Shift + K on Mac or Alt + Shift + K on Linux and Windows.

Here are some of the RStudio shortcuts:

- Insert the <- assignment operator using Option + - on Mac or Alt + - on Linux and Windows.

- Insert the pipe operator %>% using Command + Shift + M on Mac or Ctrl + Shift + M on Linux and Windows.

- Run the current line of code using Command + Enter on Mac or Control + Enter on Linux and Windows.

- Run all lines of code using Command + A + Enter on Mac or Control + A + Enter on Linux and Windows.

- Restart the current R session and start over using Command + Shift + F10 on Mac or Control + Shift + F10 on Linux and Windows.

- Comment or uncomment lines using Command + Shift + C on Mac or Control + Shift + C on Linux and Windows.

- Trying to remember a command you sent earlier? Search the command history from the console using Command + [up arrow] on Mac or Control + [up arrow] on Linux and Windows.

Quick Navigation between Windows

RStudio window panels keep important information about your project accessible. Knowing how to switch between panels without moving the mouse cursor will save time and improve your workflow. Use these shortcuts to instantly switch between panels, for example,

- Control/Ctrl + 1: Source code editor (your files)

- Control/Ctrl + 2: Console

- Control/Ctrl + 3: Help

- Control/Ctrl + 4: History

- Control/Ctrl + 5: Files

- Control/Ctrl + 6: Charts (Plots)

- Control/Ctrl + 7: Packages

- Control/Ctrl + 8: Environment

- Control/Ctrl + 9: Viewer

Console Keyboard Shortcuts

Windows and Linux	Mac	Description
Ctrl + 2	Ctrl + 2	It is used to move the cursor to the Console
Ctrl + L	Ctrl + L	It is used to clear Console
Home	Cmd + Left	It is used to move the cursor to the beginning of the line
End	Cmd + Right	It is used to move the cursor to the end of line
Up/Down	Up/Down	It is used to navigate the command history
Ctrl + Up	Cmd + Up	It is used to popup command history
Esc	Esc	It interrupts the currently executing command
Ctrl + Shift + H	Ctrl + Shift + H	It is used to change working directory

Completions (Console and Source)

Windows and Linux	Mac	Description
Attempt completion	Tab or Ctrl + Space	It is Tab or Cmd + Space
Navigate candidates	Up/Down	It is used for Up/Down
Accept selected candidate	Enter, Tab, or Right	Enter, Tab, or Right
Dismiss completion popup	Esc	Esc

Help

Windows and Linux	Mac	Description
Alt + Shift + K	Option + Shift + K	Show keyboard shortcut reference
Ctrl + Alt + F1	Ctrl + Option + F1	Search R help
Ctrl + F	Cmd + F	Find in help topic
Shift + Alt + F2	Shift + Option + F2	Previous help topic
Shift + Alt + F3	Shift + Options + F3	Next help topic
Ctrl + Shift + P, Ctrl + Alt + Shift + P (Firefox)	Cmd + Shift + P	Show command palette

Debug

Mac	Description	Windows and Linux
Shift + F9	Shift + F9	It is used for toggle Breakpoint
F10	F10	It is used to execute Next Line
Shift + F4	Shift + F4	It is used as step into Function
Shift + F7	Shift + F7	It is used to finish Function/Loop
Shift + F5	Shift + F5	It is used Continue
Shift + F8	Shift + F8	It is used to stop Debugging

Plots

Windows and Linux	Mac	Description
Ctrl + Alt + F11	Cmd + Option + F11	It shows the previous plot
Ctrl + Alt + F12	Cmd + Option + F12	It shows the next plot

Git/SVN

Windows and Linux	Mac	Description
Ctrl + Alt + D	Ctrl + Option + D	Diff active source document
Ctrl + Alt + M	Ctrl + Option + M	Commit changes
Ctrl + Up/Down	Ctrl + Up/Down	Scroll diff view
Spacebar	Spacebar	Stage/Unstaged (Git)
Enter	Return	Stage/Unstaged and move to next (Git)

Session

Windows and Linux	Mac	Description
Ctrl + Q	Cmd + Q	Quit session (desktop only)
Ctrl + Shift + F10	Cmd + Shift + F10	It restarts R session

Terminal

Windows and Linux	Mac	Description
Alt + Shift + R	Shift + Option + R	New terminal
Alt + Shift + M	Shift + Option + M	Move focus to terminal
Alt + Shift + F11	Shift + Option + F11	Previous terminal
Alt + Shift + F12	Shift + Option + F12	Next terminal

Accessibility

Windows and Linux	Mac	Description
Alt + Shift + /	Ctrl + Option + /	Toggle Screen Reader Support
Ctrl + Alt + Shift + T	Ctrl + Option + Shift + T	Toggle Tab Key Always Moves Focus
Ctrl + Alt + Shift + B	Ctrl + Option + Shift + B	Speak Text Editor Location
Alt + Shift + Y	Ctrl + Option + Y	Focus Main Toolbar
Alt + Shift + 2	Ctrl + Option + 2	Focus Console Output
F6	F6	Focus Next Pane
Shift + F6	Shift + F6	Focus Previous Pane

Main Menu (Server)

Windows and Linux	Mac	Description
Alt + Shift + F	Ctrl + Option + F	File Menu
Alt + Shift + E	Ctrl + Option + I	Edit Menu
Alt + Shift + C	Ctrl + Option + C	Code Menu
Alt + Shift + V	Ctrl + Option + V	View Menu
Alt + Shift + P	Ctrl + Option + P	Plots Menu
Alt + Shift + S	Ctrl + Option + S	Session Menu
Alt + Shift + B	Ctrl + Option + B	Build Menu
Alt+ Shift + U	Ctrl + Option + U	Debug Menu
Alt + Shift+ I	Ctrl + Option + O	Profile Menu
Alt + Shift + T	Ctrl + Option + T	Tools Menu
Alt + Shift + H	Ctrl + Option + H	Help Menu

DATA TYPES IN R

In general, when programming in any programming language, you need to use different variables to store different information.[13] Variables are nothing more than reserved memory locations for storing values. It means that when you create a variable, you reserve some space in memory.

You may want to store information of various data types such as character, wide character, integer, floating point, double floating point, Boolean, and so on. Based on the type of the variable, the operating system allocates memory and then decides what can be stored in the reserved memory.

Unlike other languages like C and Java, variables in R are not declared as any data type. R objects are assigned to variables, and the data type of the R object becomes the data type of the variable. There are many types of R objects. These are also called data structures. The most frequently used are

- Vectors

- Lists

- Matrices

- Arrays

- Factors

- Data Frames

- Table

Everything in R is an object. R has five basic atomic classes:

- logical (e.g., TRUE, FALSE)

- integer (e.g., 2L, as.integer(3))

- numeric (real or decimal) (e.g., 2, 2.0, pi)

- complex (e.g., 1 + 0i, 1 + 4i)

- character (e.g., "a," "swc")

The simplest of the objects is the vector object; there are six data types of these atomic vectors, also called the six vector classes. Other R objects are built on atomic vectors.

A variable can store various types of values such as numbers, characters, and so on. These different types of data that we can use in any code are called data types. The name of a given variable is known as its variable name. Usually, one variable stores only data belonging to a certain data type. They are so named because when the program starts it is subject to change and therefore changes from time to time.

Example:

```
x <- 123
Here, 123 is an integer data. So the data type of
variable x is integer. We can verify this by
printing the class of x variable.
> x <- 123
# print value of x variable
> print(x)
# print type of x
> print(class(x))
Output
[1] 123
> print(class(x))
[1] "numeric"
```

Here, x is a variable of the data type integer. Now, let's discuss the above data types below:

R supports three ways of assigning variables, such as

• When using the same operator, the data are copied from right to left.

• The left operator copies data from right to left.

• Using the right operator, data are copied from left to right.

Syntax:

```
#using equal to operator
variable_name = value

#using leftward operator
variable_name <- value

#using rightward operator
value -> variable_name
```

Example:

```
> # R program to illustrate
> # Initialization of variables

> # using equal to operator
> var1 = "R Programming"
> print(var1)
[1] "R Programming"

> # using leftward operator
> var2 <- "R Programming"
> print(var2)
[1] "R Programming"

> # using rightward operator
> "R Programming" -> var3
> print(var3)
[1] "R Programming"
```

Logical Data Type

The logical data type is also known as the Boolean data type.[14] It can have two values: TRUE and FALSE.

Example:

```
> bool1 <- TRUE
> bool2 <- False
> print(bool1)
> print(bool2)
[1] TRUE
[1] False
> print(class(bool1))
>print(class(bool2))
[1] "logical"
[1] "logical"
```

In the above code, Bool1 has the value of True and Bool2 has the value of FALSE.

The class() function in R is used to return the values of the class attribute of an R object. The following is the syntax of the method:

```
class (x)
```

The class() function takes the value of a parameter.

x: Represents the R object whose class attribute is to be determined.

Numeric Data Type

In R, the numeric data type represents all numbers with or without decimal values.

Example:

```
# It is floating point values
> w <- 635
> print(w)
[1] 635
> print(class(w))
[1] "numeric"
> h <- 182
> print(typeof(w))
[1] "double"
> print(h)
[1] 182
> print(class(h))
[1] "numeric"
> print(typeof(w))
[1] "double"
```

Integer Data Type

The integer data type specifies values without decimal points. We use the suffix L to specify integer data.

Example:

```
> int_var <- 168L
> print(int_var)
[1] 168
```

R supports integer data types that are the set of all integers. You can create and convert a value to an integer type using the as.integer() function. You can also use an uppercase "L" as a suffix to indicate a specific integer value.

Example:

```
# A simple R program
# to get integer data type

# here we create an integer value
> x = as.integer(5)

# print the classname of x
> print(class(x))

# print the type of x variable
print(typeof(x))

#here we declare an integer by appending an L
suffix.
y = 5L

# print the classname of y
print(class(y))

# print the type of y
print(typeof(y))
```

Complex Data Type

The complex data type is used to define purely imaginary values in R. We use the suffix i to specify the imaginary part. R supports complex data types, which are the set of all complex numbers. A complex data type is storing numbers with an imaginary component.

Example:

```
# It 2i represents imaginary part
complex_value <- 3 + 2i

# It prints class of complex_value
print(class(complex_value))

> comp_value <- 3 + 2i
> print(comp_value)
[1] 3+2i
> print(class(comp_value))
[1] "complex"
```

Character Data Type

The character type is used to specify character or string values in a variable. A string is a set of characters. Example, 'A' is a single character and "R Programming" is a string. You can use single quotes ' ' or double quotes " " to represent strings. In general, we use:

- ' ' for character variables

- " " for string variables

```
# It creates a string variable
> language <- "R Programming"
> print(class(language))
[1] "character"
> print(language)
[1] "R Programming"

# It creates a character variable
> my_char <- 'A'
> print(class(my_char))
[1] "character"
> print(my_char)
[1] "A"
```

Raw Data Type

The raw data type specifies values as raw bytes. You can use the same methods to convert character data types to a raw data type and vice versa.

- charToRaw() - It converts character data to raw data

- rawToChar() - It converts raw data to character data

Here is the basic data types table given below:

Basic data types	Values
Numeric	It is set of all real numbers
Integer	It is set of all integers, Z
Logical	TRUE and FALSE
Complex	It is set of complex numbers
Character	"a", "b", "c", ..., "@", "#", "$", ..., "1", "2", ...etc.

Here is the table of R data type:

Logical True, False It is a special data type for data with two possible values which can be construed as true/false			
Numeric 2,312,112,5432 It is the decimal value is called numeric in R, and it is the default computational data type			
Integer 3L, 66L, 2346L Here, it is L tells R to store the value as an integer			
Complex Z=1+2i, t=7+3i A complex value is defined as the pure imaginary value i			
Character 'a', "good", "TRUE", '35.4' In R programming, a character is used to represent string values. We convert the objects into character values with the help of .character() function			
Raw A data type is used to hold raw bytes			

Find Data Type of an Object

To find the data type of any object, use the class() function.[15] The syntax is you need to pass the object as a parameter to the function class() to find the data type of an object.

Here is the syntax of the class, class(object)

Example:

```
> # A simple R program
> # to find data type of an object
>
> # Logical
> print(class(FALSE))
[1] "logical"
>
```

```
> # Integer
> print(class(2L))
[1] "integer"
>
> # Numeric
> print(class(10.5))
[1] "numeric"
>
> # Complex
> print(class(2+2i))
[1] "complex"
>
> # Character
> print(class("19-07-2020"))
[1] "character"
```

Type Verification

To do so, you must use the "is" prefix before the data type as a statement. The syntax for this is is.data_type() of the object you need to validate.

Syntax:

```
is.data_type(object)
```

Example:

```
> # A simple R program
> # check if an object is of certain data type
> # Logical
> print(is.logical(TRUE))
[1] TRUE
> # Integer
> print(is.integer(3L))
[1] TRUE
> # Numeric
> print(is.numeric(10.5))
[1] TRUE
> # Complex
> print(is.complex(1+2i))
[1] TRUE
> # Character
> print(is.character("12-04-2020"))
```

```
[1]  TRUE
> print(is.integer("a"))
[1]  FALSE
> print(is.numeric(2+3i))
[1]  FALSE
```

Convert an Object's Data Type to Another

This task is where you can change or convert the data type of an object to another. You must use the "as" prefix to perform this action, before the data type as a command. The syntax for this is as.data_type() of the object you want to force.

Syntax:

```
as.data_type(object)
```

Example:

```
> # R program
> # convert data type from an object to another
> # Logical
> print(as.numeric(TRUE))
[1]  1
> # Integer
> print(as.complex(3L))
[1]  3+0i
> # Numeric
> print(as.logical(10.5))
[1]  TRUE
> # Complex
> print(as.character(1+2i))
[1]  "1+2i"
> # Can't possible
> print(as.numeric("12-04-2020"))
[1]  NA
```

NOMENCLATURE OF R VARIABLES

When naming a variable, keep the following rules in mind:[16]

- A valid variable name is a combination of alphabets, numbers, periods (.), and underscores (_). Example: var.1_ is valid

- Other than the dot and underscore operators, no other special characters are allowed. Example: var$1 or var#1 is invalid

- Variables can start with alphabets or periods. Example:. var or var is valid

- A variable should not start with numbers or an underscore. Example: 22var or _var is invalid

- If a variable starts with a dot (.), then the next thing after the dot cannot be a number. Example: .3var is invalid

IMPORTANT METHODS FOR VARIABLES

R language provides some useful methods for performing operations on variables. Below methods are used to determine the data type of a variable, find a variable, delete a variable, and so on. The following are some of the methods used to work with variables.

class() Function

This built-in function is used to find the data type of the variable that is provided to it. The variable to be checked is passed as an argument and returns the data type.

Syntax:

```
class(variable)
> var1 = "hello"
> # you can write variable as
> # var1 <- "hello"
> print(class(var1))
[1] "character"
```

ls() Function

This function is used to know all the variables present in the workspace. This is generally useful when working with a large number of variables at once and helps prevent overwriting any of them.

Syntax:

```
ls()
```

Example:

```
> #first variable
> var = "bool1"

> #second variable
> var = "value"

> # third variable
> var = "int_val"

> # fourth variable
> var = "language"

> # fifth variable
> var = "my_char"

> # using equal to operator
> var1 = "hello"

> # using equal to operator
> var1 = "hello"
> # using leftward operator
> var2 <- "hello"
> # using rightward operator
> "hello" -> var3
> print(ls())
 [1] "bool1"      "value"
 [3] "int_var"    "language"
 [5]    "my_char"  "var1"
 [7]          "var2"      "var3"
```

rm() Function

This is a built-in function to remove an unwanted variable in your work-space. This helps clean up the memory space allocated to certain variables that are not being used, thus making more room for others. The name of the variable to be deleted is passed as an argument.

Syntax:

```
rm(variable)
```

Example:

```
> # using equal to operator
> var1 = "R programming"
> # using leftward operator
> var2 <- "R programming"
> # using rightward operator
> "R programming" -> var3
> # Removing variable
> rm(var3)
> print(var3)
Error in print(var3) : the object 'var3' not found
Execution halted
```

Scope of Variable in R

In R, variables are some containers for storing data values.[17] They are references or pointers to an object in memory, meaning that whenever a variable is assigned to an instance, it maps to the instance. A variable can store a vector, a group of vectors, or a combination of many R objects.

```
> # R program to demonstrate
> # variable assignment
> # The assignment using equal operator
> var1 = c(00, 11, 22, 33)
> print(var1)
[1]  0 11 22 33
> # The assignment using leftward operator
> var2 <- c("Python Programming", "R Programming")
> print(var2)
[1] "Python Programming"
[2] "R Programming"
> # A Vector Assignment
> a = c(1, 2, 3, 4)
> print(a)
[1] 1 2 3 4
> b = c("ABC", "BCD", "CDE", "FGH")
> print(b)
[1] "ABC" "BCD" "CDE" "FGH"
> # A group of vectors Assignment using list
> c = list(a, b)
> print(c)
[[1]]
```

```
[1] 1 2 3 4
[[2]]
  [1] "ABC" "BCD" "CDE" "FGH"
```

VARIABLE NAMING CONVENTION

- A variable name in R must be alphanumeric characters with the exception of underscore('_') and period('.'), which are special characters that can be used in variable names.

- A variable name must always start with an alphabet.

- Other special characters like '!', '@', '#', '$' are not allowed in variable names.

The Scope of a Variable

The place where we can find the variable and also access it when needed is called the scope of the variable. There are mainly two types of variable ranges:

- Global variables: The global variables are those that exist during program execution. They can be changed and accessed from any part of the program.

They are useable for the entire duration of the program. They are declared anywhere in the code outside of any functions or blocks.

Declaring global variables: The global variables are declared outside of all functions and blocks. They can be accessed from any part of the program.

Example:

```
> # R program to get to know the
> # usage of global variables
> # global variable
> global_variable = 5
> # Here global variable accessed from
> # within a function
> display = function(){
+     print(global)
+ }
```

```
> display()
[1] 10
> # changing value of global variable
> global_var
```

Explanation: The "global_variable" variable is declared at the top of the program outside of any functions; it is a global variable and can access or update from anywhere in the program.

- <Local variables: The local variables are those variables that exist only in a particular part of the program such as a function and are freed when the function call ends.

This variable does not exist outside the block in which they are declared, that is, they cannot be accessed or used outside that block. Declaration of local variables: Local variables are declared inside a block.

Example:

```
> # R program to illustrate
> # usage of local variables
> func = function(){
+        # this variable is local to the
+        # function func() and cannot be
+        # accessed outside this function
+        user_age = 18
+ }
> print(user_age)
Error in print(age) : the object 'age' not found
```

The above program shows an error that "object 'age' not found." The age variable was declared in the "func()" function; so, it is local to that function and not visible to any part of the program outside of that function.

To get the output of the above function, we need to display the value of the age variable only from the "func()" function.

Example:

```
> # R program to illustrate
> # usage of local variables
> func = function(){
```

```
+        # this variable is local to the
+        # function func() and cannot be
+        # accessed outside this function
+        user_age = 18
+        print(user_age)
+ }
> cat(" User age is:\n")
 User age is:
> func()
[1] 18
```

How to Access Local Variable Globally
Global variables can be used from anywhere in the code, unlike local variables, which have scope limited to the block of code in which they are created.

Example:

```
> f = function() {
+        # a is a local variable here
+        a <-1
+ }
> f()
>
> # Can't access outside the function
> a   # This'll give error
Error in print(a) : object 'a' not found
```

In the code, we can see that we cannot access the variable "a" outside the function because it is assigned by the assignment operator (<-), which makes "a" a local variable. The super assignment operator (<<-) is used to assign global variables.

How Does the Super Assignment Operator Work?
When this operator is used within a function, it looks for a variable within the parent environment; if not found, it continues searching up the next level until it reaches the global environment. If the variable is not found, it is created and assigned at the global level.

Example:

```
> # R program to find the scope of variables
> outer_function = function(){
```

```
+        inner_function = function(){
+             # this "<<-" operator here
+             # it makes a as global variable
+             a <<- 100
+             print(a)
+        }
+        inner_function()
+        print(a)
+ }
> outer_function()
[1] 100
[1] 100
```

When it runs the statement "a <<- 100" within inner_function(), it looks for the variable "a" in external_function. Since "a" is not defined in the global environment either, it is created and assigned here, which is now referenced and printed within both inner_function() and external_function().

R Programming Environment

When a starts a new project in R, the R backend system creates a new environment for the objects (variables) created during that session. The environment is called the global environment. The global environment is not the root of the environment tree. It is actually the last environment in the environment chain on the search path.

Example:

```
> a <- 2
> b <- 5
> f <- function(x) x<- 0
> ls()
 [1] "a"
 [2] "b"
 [3] "c"
 [4] "display"
 [5] "f"
 [6] "func"
 [7] "global"
```

```
 [8] "global_variable"
 [9] "outer_function"
[10] "var1"
[11] "var2"
[12] "x"
> environment()
<environment: R_GlobalEnv>
> .GlobalEnv
<environment: R_GlobalEnv>
```

An environment can be understood as a set of objects (functions, variables, etc.). An environment is created when we first start the R interpreter. Any variable we define is now in that environment.

The top-level environment available to us on the R command line is a global environment called R_GlobalEnv. The global environment can also be referred to as .GlobalEnv in R codes.

Using the ls() function, we can show what variables and functions are defined in the current environment. Additionally, we can use the environment() function to get the current environment.

DATA STRUCTURE IN R

A data structure is a way of well-organizing data in a computer so that it can be used efficiently.[18] The goal is to reduce the spatial and temporal complexity of various tasks. Data structures in R programming are tools for storing multiple values.

Basic R data structures are often organized by their dimensionality (1D, 2D, or nD) and by whether they are all elements must be of the same type (homogeneous) or elements are often of different types (heterogeneous). It creates six data types that are most often used in data analysis.

The most essential data structures used in R include:

- Vectors

- Lists

- Data frames

- Matrices

- Arrays

- Factors

GETTING DATA IN R

There are several main functions for reading data into R:[19]

- read.table() and read.csv() are two popular functions used for reading tabular data into R reading tabular data

- readLines()function, to read the lines of a text file

- source()function, for reading R code files (inversion to dump)

- dget()function, for reading R code files (inversion of dput)

- load() function, for reading in saved workspaces

- unserialize() function, for reading individual R objects in binary form.

There are many R packages that have been developed to read in all types of other datasets and may need to resort to one of these packages if you work in a certain area.

There are various functions exist for writing data to files:

- write.table(), to write table data to text files (i.e., CSV) or concatenation

- writeLines(), to write line-by-line character data to a file or connection

- dump(), to dump the textual representation of multiple R objects

- put(), to output a textual representation of an R object

- save(), to save any number of R objects in binary format (possibly compressed) to a file.

- Serialize(), to convert an R object to binary format for output to a connection (or file).

CHAPTER SUMMARY

We did the installation of RStudio application in this chapter. In R, we keep our data in a variable and that variable can be used further in decision-making statements and loops.

NOTES

1. RStudio User Interface – https://grunwaldlab.github.io/analysis_of_microbiome_community_data_in_r/00--intro_to_rstudio.html, accessed on September 24, 2022.

2. RStudio Console – https://grunwaldlab.github.io/analysis_of_microbiome_ community_data_in_r/00--intro_to_rstudio.html, accessed on September 24, 2022.

3. Tip and Trick – https://support.rstudio.com/hc/en-us/articles/200404846- Working-in-the-Console-in-the-RStudio-IDE, accessed on September 24, 2022.

4. Keyboard Shortcuts – https://support.rstudio.com/hc/en-us/articles/200711853, accessed on September 24, 2022.

5. Variable in R – https://www.w3schools.com/r/r_variables.asp, accessed on September 24, 2022.

6. Comments in R – https://www.tutorialspoint.com/r/r_basic_syntax.htm, accessed on September 24, 2022.

7. Keyword in R – https://www.datamentor.io/r-programming/reserved- words/, accessed on September 24, 2022.

8. Function – https://www.javatpoint.com/r-functions, accessed on September 26, 2022.

9. Break in R – https://www.javatpoint.com/r-break-statement#:~:text=In%20 the%20R%20language%2C%20the,control%20the%20program%20execu- tion%20flow, accessed on September 24, 2022.

10. Null – https://r-lang.com/null-in-r/, accessed on September 26, 2022.

11. NA Character in R – https://www.njtierney.com/post/2020/09/17/missing- flavour/, accessed on September 27, 2022.

12. Keyboard Shortcut – https://www.dataquest.io/blog/rstudio-tips-tricks- shortcuts/, accessed on September 26, 2022.

13. Data Types – https://www.tutorialspoint.com/r/r_data_types.htm, accessed on September 27, 2022.

14. Data types – https://www.geeksforgeeks.org/r-data-types/, accessed on September 27, 2022.

15. Data Types – https://www.geeksforgeeks.org/r-data-types/, accessed on September 27, 2022.

16. Data Type Nomenclature – https://www.geeksforgeeks.org/r-variables/, accessed on September 27, 2022.

17. Scope of Variables – https://www.geeksforgeeks.org/scope-of-variable-in-r/, accessed on September 28, 2022.

18. Data Structures – https://www.geeksforgeeks.org/data-structures-in-r- programming/, accessed on September 28, 2022.

19. Data in R – https://makemeanalyst.com/r-programming/reading-and- writing-data-to-and-from-r/, accessed on September 28, 2022.

## CHAPTER 3

Variables and Operators

IN THIS CHAPTER

> ➤ Introduction

> ➤ Constants in R

> ➤ Basic Operations

> ➤ Operators

In the last chapter, we had a discussion on data types which we will use in the subsequent chapters; before data types, we also covered the RStudio application basic overview with some keyboard shortcuts to make work easy.

INTRODUCTION

Variables and constants are the basic units that are used to develop a program.[1] Almost all programming languages provide the function of using variables and constants. In this chapter, you will know the concepts of variables, constants with some basic methods of using vectors in the R program. Variables are used to store data, the value of which can be changed according to our needs. The unique name of a given variable (both functions and objects) is an identifier.

Rules for Writing Identifiers in R

- Identifiers can be a combination of letters, numbers, periods (.), and underscores (_).

DOI: 10.1201/9781003358480-3

- Identifiers must start with a letter or a period. If it starts with a dot, it can't be followed by a number.

- The reserved words in R can't be used as identifiers.

Constants in R

Constants as the name suggests, are basically the entities whose value cannot be changed.[2] The basic types of constants are numeric and character. Here you will know good practice of variable.

- The old versions of R used the underscore (_) as the assignment operator. Thus, the dot (.) was widely used in multi-word variable names.

- Now, current versions of R support the underscore as a valid identifier, but it is good practice to use periods as word separators.

For example, a.variable.name is always preferred over a_variable_name, or alternatively, we can use camel size as aVariableName.

Valid Identifiers in R

Here is an example of valid identifiers in R:

```
# Here is the normal lowercase text
> total <- 12
# Here is the normal capitalize text
> Sum <- 12 + 12
# Here is the text with dot(.)
> .fine.with.dot <- 12
# Here is the text with dot(.)
> this_is_underscore <- 12
# Here is the text with uppercase
>   TEXT <- 12
> print(total)
[1] 12
> print(Sum)
[1] 24
> print(.fine.with.dot)
[1] 12
> print(this_is_underscore)
[1] 12
> print(TEXT)
[1] 12
```

Invalid Identifiers in R

Here is an example of invalid identifiers in R:

```
> total@1
Error: unexpected numeric constant in "total@1"
> 5sum <- 12
Error: unexpected symbol in "5sum"
> _fine <- 12
Error: unexpected symbol in "_fine"
> TRUE <- 12
Error in TRUE <- 12: invalid (do_set) left-hand side
to assignment
```

Numerical Constants

All numbers fall into this category. They can be of data-type integer, double or complex. They can be checked using the typeof() function. Numeric constants followed by L are considered integer, and constants followed by i are considered complex.

Example:

```
> typeof(6)
[1] "double"

> typeof(4L)
[1] "integer"

> typeof(34.34)
[1] "double"

> print(typeof(2 + 3i))
[1] "complex"
```

Numeric constants preceded by 0x or 0X are interpreted as hexadecimal numbers.

Here is an example of the hexadecimal number:

```
> 0XF
[1] 15
> 0xff
[1] 255
```

```
> 0xf
[1] 15
> 0xfff + 0xf
[1] 4110
> 0xff + 0xf
[1] 270
```

Character Constants

Character constants can represent using either single quotes (') or double quotes (") as delimiters.

Here is an example of the character:

```
> ' R programming '
[1] " R programming"
> typeof("R programming")
[1] "character"
> a <- "R programming"
> print(a)
[1] "R programming"
```

Constants: Built-in Constants

R has a tiny number of built-in constants.[3] The following constants are available:

- LETTERS: there are 26 uppercase letters of the Roman alphabet.

- letters: there are 26 lowercase letters of the Roman alphabet.

- month.abb: there are three-letter abbreviations for the English month names.

- month.name: there are English names for the months of the year.

- pi: It is the ratio of the whole circumference of a circle to its diameter.

These are the commonly used built-in constants. Here is an example of each of them.

```
> # Here is the abbreviation of letters in capital
> LETTERS
 [1] "A" "B" 'C' "D" "E" "F" "G" "H"
 [9] "I" "J" 'K' "L" "M" "N" "O" "P"
[17] "Q" "R" 'S' "T" "U" "V" "W" "X"
[25] "Y" "Z"
```

```
> # Here is the abbreviation of letters in small
> letters
 [1] "a" "b" 'c' "d" "e" "f" "g" "h"
 [9] "i" "j" "k" 'l' "m" "n" "o" "p"
[17] "q" "r" 's' "t" "u" "v" "w" "x"
[25] "y" "z"

> # Here is the abbreviation of Month Full Name
> month.name
 [1] "January"   "February"
 [3] "March"     "April"
 [5] "May"       "June"
 [7] "July"      "August"
 [9] "September" "October"
[11] "November"  "December"

> # Here is the abbreviation of Month Short Name
> month.abb
 [1] "Jan" "Feb" "Mar" "Apr" "May"
 [6] "Jun" "Jul" "Aug" "Sep" "Oct"
[11] "Nov" "Dec"

# Here is the value of pi
> pi
[1] 3.141593
```

BASIC OPERATIONS IN R

<-

As a programming language, R has variables (or "objects") to which "things" are assigned (values, functions, other variables). These assignments use the <- operator. For example, to assign the value 15 to a variable named x, we write:

x <- 15

Others might get confused why it's not x = 15. In R = is rarely used and can often be used instead of <-, but this is generally not a good idea. When we define functions and when we call functions, we use the = sign to assign or pass values to the function.

We now have an x object that has a value of 15 in our environment. In fact, you can look at the Environment pane at the top right of RStudio.

R Is Case-Sensitive

R is case-sensitive, so gaData is a completely different object from gadata (and x is different from X). Depending on how well you write and how carefully you pay attention to detail, this is an easy way to get tripped up. One more time: R IS CASE-SENSITIVE.

Function

Functions are the foundation of any programming language, and R is no different. There are three main sources of functions.

Defined in Base R

There are many, and we'll touch on some common ones in this section. Consider the following code:

```
> sum(c(1, 2, 3, 4, 5))
[1] 15
```

This code uses two functions:

- c() – This is a "combine" function. This example creates a "vector" of numeric values.

- sum() – this is just like the SUM() or sum() function in another programming language.

R OPERATORS

An operator is a symbol that tells the compiler to perform specific mathematical or logical manipulations.[4] The R language is rich in built-in operators and provides the following types of operators.

Types of Operators

We have the following types of operators in R programming:

- Arithmetic operators

- Relational operators

- Logical operators

- Assignment operators

- Miscellaneous operators

We shall learn about these operators in brief with example R programs.

R ARITHMETIC OPERATORS

Operator	Explanation	Usage
+	It is used for addition of two operands	a + b
−	It is used for subtraction of the second operand from first	a − b
*	It is used for multiplication of two operands	a * b
/	It is used for division of the first operand with second	a / b
%%	It is used to get remainder from division of the first operand with second	A %% b
%/%	It is used to get quotient from division of the first operand with second	A %/% b
^	It is used for the first operand raised to the power of the second operand	a ^ b
%*%	Matrix multiplication	
%o%	Outer product	
%x%	Kronecker product	

An example for each arithmetic operator on values is provided in the following code:

```
# Here we assign the value to a and b variable,
> a <- 7.5
> b <- 2

#the result is given below,
> print ( a+b )    #addition
[1] 9.5
> print ( a-b )    #subtraction
[1] 5.5
> print ( a * b )    #multiplication
[1] 15
> print ( a / b )    #Division
[1] 3.75
> print ( a %%b )  #Reminder
[1] 1.5
> print ( a%/%b )  #Quotient
[1] 3
> print ( a ^ b )    #Power of
[1] 56.25
```

Another example of arithmetic vector is given below:

```
# R Operators - R Arithmetic Operators Example for
vectors
> # R Operators - R Arithmetic Operators Example for
vectors
> # Here we assigning vector value to variable a and
b,
> a <- c(8, 9, 6)
> b <- c(2, 4, 5)
> print ( a+b )    # addition of two vector
[1] 10 13 11
> print ( a-b )    # subtraction of two vector
[1] 6 5 1
> print ( a*b )    # multiplication of two vector
[1] 16 36 30
> print ( a/b )    # division of two vector
[1] 4.00 2.25 1.20
> print ( a%%b )   # reminder of two vector
[1] 0 1 1
> print ( a%/%b )  # quotient of two vector
[1] 4 2 1
> print ( a^b )    # power of two vector
```

MATRIX OPERATIONS IN R

There are several matrix operations that you can perform in R.[5] These include addition, subtraction, and multiplication, calculating the power, order, determinant, diagonal, eigenvalues and eigenvectors, transposing, and factoring a matrix by various methods.

Addition and Subtraction

The most basic matrix operations are addition and subtraction. In these examples, we are going to use the square matrices of the following block of code:

```
# Here is the matrix A
> A <- matrix(c(10, 8,
+                5, 12), ncol = 2, byrow = TRUE)

# Here is the matrix B
> B <- matrix(c(5, 3,
+                15, 6), ncol = 2, byrow = TRUE)
```

```
# The result is here,
> print(B)
     [,1] [,2]
[1,]    5    3
[2,]   15    6
> print(A)
     [,1] [,2]
[1,]   10    8
[2,]    5   12

> is.matrix(A)
[1] TRUE
> is.vector(A)
[1] FALSE
> is.matrix(B)
[1] TRUE
> is.vector(B)
[1] FALSE
> dim(A)
[1] 2 2
> dim (B)
[1] 2 2
```

Both these matrices are of the same dimensions. You can check the dimension number of rows and columns, respectively, of a matrix with the dim function. Here is an example:

```
> A <- matrix(c(18, 18,
+                5, 12, 10, 5), ncol = 3, byrow = TRUE)
> B <- matrix(c(5, 3,
+                15, 6, 1, 8), ncol = 3, byrow = TRUE)
> print(A)
     [,1] [,2] [,3]
[1,]   18   18    5
[2,]   12   10    5
> print(B)
     [,1] [,2] [,3]
[1,]    5    3   15
[2,]    6    1    8
> dim(A)
[1] 2 3
> dim(B)
[1] 2 3
```

You can also apply arithmetic operator on the matrix such as addition, subtraction, multiplication, division, and so on.

1. The + operator allows you to compute an element-wise sum of the two matrices such as: A + B

Example:

```
> A <- matrix(c(18, 18,
+                5, 12, 10, 5), ncol = 3, byrow =
TRUE)
>
> B <- matrix(c(5, 3,
+                15, 6, 1, 8), ncol = 3, byrow =
TRUE)
> dim(B)
[1] 2 3
> A + B
      [,1] [,2] [,3]
[1,]   23   21   20
[2,]   18   11   13
```

2. The - operator will allow you to subtract two matrices such as: A - B

Example:

```
> A <- matrix(c(18, 18,
+                5, 12, 10, 5), ncol = 3, byrow =
TRUE)
>
> B <- matrix(c(5, 3,
+                15, 6, 1, 8), ncol = 3, byrow =
TRUE)
> dim(B)
[1] 2 3

> A - B
      [,1] [,2] [,3]
[1,]   13   15  -10
[2,]    6    9   -3
```

TRANSPOSE A MATRIX IN R

If you want to find the transpose of a matrix in R, you only need to use the
t function as follows:

```
> A <- matrix(c(18, 18,
+                 5, 12, 10, 5), ncol = 3, byrow = TRUE)

> B <- matrix(c(5, 3,
+                 15, 6, 1, 8), ncol = 3, byrow = TRUE)

> print(A)
     [,1] [,2] [,3]
[1,]   18   18    5
[2,]   12   10    5

> print(B)
     [,1] [,2] [,3]
[1,]    5    3   15
[2,]    6    1    8

# Here is the
> t(A)
     [,1] [,2]
[1,]   18   12
[2,]   18   10
[3,]    5    5

> t(B)
     [,1] [,2]
[1,]    5    6
[2,]    3    1
[3,]   15    8
```

Matrix Multiplication in R

There are various different ways of matrix multiplication such as by a scalar, element-wise multiplication, matricial multiplication, exterior, and
Kronecker product. Let's discuss each of them in brief.

Multiplication by a Scalar

If you want to multiply or divide a matrix by a scalar, you can make use of
the " * " or " / " operators, respectively:

Here is an example of multiple.

```
> Here we assign three by three matrix to A and B
variables,
> A <- matrix(c(18, 18,
+                    5, 12, 10, 5), ncol = 3, byrow = TRUE)
> B <- matrix(c(5, 3,
+                    15, 6, 1, 8), ncol = 3, byrow = TRUE)
> print(A)
     [,1] [,2] [,3]
[1,]   18   18    5
[2,]   12   10    5
> print(B)
     [,1] [,2] [,3]
[1,]    5    3   15
[2,]    6    1    8
> # Here we multiple matrix by a scalar
> 2 * A
     [,1] [,2] [,3]
[1,]   36   36   10
[2,]   24   20   10
> 2 * B
     [,1] [,2] [,3]
[1,]   10    6   30
[2,]   12    2   16
```

Here is an example of divide /.

```
> # Here we assign three by three matrix to A and B
variables,
> A <- matrix(c(18, 18, 5, 12, 10, 5), ncol = 3, byrow
= TRUE)
> B <- matrix(c(5, 3, 15, 6, 1, 8), ncol = 3, byrow =
TRUE)
> print(A)
     [,1] [,2] [,3]
[1,]   18   18    5
[2,]   12   10    5
> print(B)
     [,1] [,2] [,3]
[1,]    5    3   15
[2,]    6    1    8
>
> # Here we divide matrix by a scalar
```

```
> 2 * A
     [,1] [,2] [,3]
[1,]   36   36   10
[2,]   24   20   10
>
> 2 / B
           [,1]         [,2]         [,3]
[1,]  0.4000000  0.6666667  0.1333333
[2,]  0.3333333  2.0000000  0.2500000
```

Element-Wise Multiplication

The element-wise multiplication of two matrices of the same dimensions can also be calculated with the * operator.

Syntax:

```
A * B
```

Example:

```
> A <- matrix(c(18, 18,
+                 5, 12, 10, 5), ncol = 3, byrow
= TRUE)
>
> B <- matrix(c(5, 3,
+                 15, 6, 1, 8), ncol = 3, byrow
= TRUE)
> A * B
     [,1] [,2] [,3]
[1,]   90   54   75
[2,]   72   10   40
```

Matrix Multiplication in R

Matrix multiplication is the most useful matrix operation.[6] A matrix in R can be created using the matrix() function and this function takes an input vector, nrow, ncol, byrow, and dimnames as arguments. In R, matrix multiplication can be performed using the %*% operator.

Example:

```
> data <- c(1, 2, 3, 0, 1, 2, 0, 0, 1)
> A <- matrix(data, nrow = 3, ncol = 3)
```

```
> data <- c(0, 1, 1, 1, 0, 3, 1, 3, 3)
> B <- matrix(data, nrow = 3, ncol = 3)
> AB <- A %*% B
> print("Matrix A")
[1] "Matrix A"
> print(A)
     [,1] [,2] [,3]
[1,]   1    0    0
[2,]   2    1    0
[3,]   3    2    1
> print("Matrix B")
[1] "Matrix B"
> print(B)
     [,1] [,2] [,3]
[1,]   0    1    1
[2,]   1    0    3
[3,]   1    3    3
> print("Matrix Multiplication Result")
[1] "Matrix Multiplication Result"
> print(AB)
     [,1] [,2] [,3]
[1,]   0    1    1
[2,]   1    2    5
[3,]   3    6   12
```

Matrix Multiplication Using Three Matrix

Here is an example of matrix multiplication using three matrix.

```
> data <- c(1, 2, 3, 0, 1, 2, 0, 0, 1)
> A <- matrix(data, nrow = 3, ncol = 3)
> data <- c(0, 1, 1, 1, 0, 3, 1, 3, 3)
> B <- matrix(data, nrow = 3, ncol = 3)
> AB <- A %*% B
> print("Matrix A")
[1] "Matrix A"
> print(A)
     [,1] [,2] [,3]
[1,]   1    0    0
[2,]   2    1    0
[3,]   3    2    1
> print("Matrix B")
[1] "Matrix B"
```

```
> print(B)
     [,1] [,2] [,3]
[1,]    0    1    1
[2,]    1    0    3
[3,]    1    3    3
> print("Matrix Multiplication Result")
[1] "Matrix Multiplication Result"
> print(AB)
     [,1] [,2] [,3]
[1,]    0    1    1
[2,]    1    2    5
[3,]    3    6   12
```

MATRIX CROSS-PRODUCT

If you want to calculate the matrix product of a matrix and a transpose or other, you can type t(A) %*% B or A %*% t(B). A, B is the name of the matrix. However, in R, it is good to use the crossprod and tcrossprod functions.

Syntax of crossprod:

(A, B)

Example:

```
> data <- c(1, 2, 3, 0, 1, 2, 0, 0, 1)
> A <- matrix(data, nrow = 3, ncol = 3)
> data <- c(0, 1, 1, 1, 0, 3, 1, 3, 3)
> B <- matrix(data, nrow = 3, ncol = 3)
 > AB <- t(A) %*% B
 > print("Matrix A")
[1] "Matrix A"
> print(A)
     [,1] [,2] [,3]
[1,]    1    0    0
[2,]    2    1    0
[3,]    3    2    1
> print("Matrix B")
[1] "Matrix B"
> print(B)
     [,1] [,2] [,3]
[1,]    0    1    1
```

```
[2,]    1    0    3
[3,]    1    3    3
> print("Matrix Crossprod Result")
[1] "Matrix Crossprod Result"
> print(AB)
     [,1] [,2] [,3]
[1,]    5   10   16
[2,]    3    6    9
[3,]    1    3    3
```

Syntax of tcrossprod:

```
(A, B)
```

Example:

```
> data <- c(1, 2, 3, 0, 1, 2, 0, 0, 1)
> A <- matrix(data, nrow = 3, ncol = 3)
> data <- c(0, 1, 1, 1, 0, 3, 1, 3, 3)
> B <- matrix(data, nrow = 3, ncol = 3)
> AB <- A %*% t(B)
> print("Matrix A")
[1] "Matrix A"
> print(A)
     [,1] [,2] [,3]
[1,]    1    0    0
[2,]    2    1    0
[3,]    3    2    1
> print("Matrix B")
[1] "Matrix B"
> print(B)
     [,1] [,2] [,3]
[1,]    0    1    1
[2,]    1    0    3
[3,]    1    3    3
> print("Matrix tcrossprod Result")
[1] "Matrix tcrossprod Result"
> print(AB)
     [,1] [,2] [,3]
[1,]    0    1    1
[2,]    1    2    5
[3,]    3    6   12
```

Exterior Product

Same as the matricial multiplication, in R, you can also compute the exterior product of 2 matrices with the %o% operator. This operator is a short-code for the default outer function.

Syntax:

```
A %o% B
```

Equivalent to:

```
outer(A, B, FUN = "*")
```

Here is an example of 3*3 matrices.

```
> data <- c(1, 2, 3, 0, 1, 2, 0, 0, 1)
> A <- matrix(data, nrow = 3, ncol = 3)
> data <- c(0, 1, 1, 1, 0, 3, 1, 3, 3)
> B <- matrix(data, nrow = 3, ncol = 3)
> AB <- A %o% B
> print("Matrix A")
[1] "Matrix A"
> print(A)
      [,1] [,2] [,3]
[1,]    1    0    0
[2,]    2    1    0
[3,]    3    2    1
> print("Matrix B")
[1] "Matrix B"
> print(B)
      [,1] [,2] [,3]
[1,]    0    1    1
[2,]    1    0    3
[3,]    1    3    3
> print("Matrix tcrossprod Result")
[1] "Matrix tcrossprod Result"
> print(AB)
, , 1, 1

      [,1] [,2] [,3]
[1,]    0    0    0
[2,]    0    0    0
[3,]    0    0    0
```

```
, ,   2, 1

      [,1] [,2] [,3]
[1,]    1    0    0
[2,]    2    1    0
[3,]    3    2    1

, ,   3, 1

      [,1] [,2] [,3]
[1,]    1    0    0
[2,]    2    1    0
[3,]    3    2    1

, ,   1, 2

      [,1] [,2] [,3]
[1,]    1    0    0
[2,]    2    1    0
[3,]    3    2    1

, ,   2, 2

      [,1] [,2] [,3]
[1,]    0    0    0
[2,]    0    0    0
[3,]    0    0    0

, ,   3, 2

      [,1] [,2] [,3]
[1,]    3    0    0
[2,]    6    3    0
[3,]    9    6    3

, ,   1, 3

      [,1] [,2] [,3]
[1,]    1    0    0
[2,]    2    1    0
[3,]    3    2    1

, ,   2, 3
```

```
        [,1]  [,2]  [,3]
[1,]      3     0     0
[2,]      6     3     0
[3,]      9     6     3

, ,   3, 3

        [,1]  [,2]  [,3]
[1,]      3     0     0
[2,]      6     3     0
[3,]      9     6     3
```

Kronecker Product

The Kronecker product of two matrices AA and BB, denoted by A \otimes BA⊗B, is the last type of matricial product; we are going to review this. The calculation can be achieved with the %x% operator.

Syntax:

```
A %x% B
```

Example:

```
> data <- c(1, 2, 3, 0, 1, 2, 0, 0, 1)
> A <- matrix(data, nrow = 3, ncol = 3)
> data <- c(0, 1, 1, 1, 0, 3, 1, 3, 3)
> B <- matrix(data, nrow = 3, ncol = 3)
> AB <- A %x% B
> print("Matrix A")
[1] "Matrix A"
> print(A)
        [,1]  [,2]  [,3]
[1,]      1     0     0
[2,]      2     1     0
[3,]      3     2     1
> print("Matrix B")
[1] "Matrix B"
> print(B)
        [,1]  [,2]  [,3]
[1,]      0     1     1
[2,]      1     0     3
[3,]      1     3     3
```

```
> print("Matrix KroneckerResult")
[1] "Matrix KroneckerResult"
> print(AB)
      [,1] [,2] [,3] [,4] [,5] [,6]
[1,]    0    1    1    0    0    0
[2,]    1    0    3    0    0    0
[3,]    1    3    3    0    0    0
[4,]    0    2    2    0    1    1
[5,]    2    0    6    1    0    3
[6,]    2    6    6    1    3    3
[7,]    0    3    3    0    2    2
[8,]    3    0    9    2    0    6
[9,]    3    9    9    2    6    6
      [,7] [,8] [,9]
[1,]    0    0    0
[2,]    0    0    0
[3,]    0    0    0
[4,]    0    0    0
[5,]    0    0    0
[6,]    0    0    0
[7,]    0    1    1
[8,]    1    0    3
[9,]    1    3    3
```

POWER OF A MATRIX IN R

The power of a matrix cannot be found directly in R because there is no built-in function in the base to calculate the power of a matrix, so you will provide two different alternatives. So, we can use % ^ % from expm package for this purpose. First, we install the exam package, then load it and use % ^ % i your code. For example, suppose you have a matrix called A, and we want to find raising M to the power of 2, then can be done as – M %^%2.

When we run the command in the console, that is, "install.package ("expm")," the output will be same as given below:

```
> install.packages("expm")
WARNING: The Rtools is required to build R packages
but is not current install. Please download and
installed the appropriate version of Rtools before
proceeding:
http://cran.rstudio.com/bin/windows/Rtools/
Installing package into 'C:/Users/Dell/AppData/
Local/R/win-library/4.2'
```

```
(as 'lib' is unspecified)
trying URL 'http://cran.rstudio.com/bin/windows/
contrib/4.2/expm_0.999-6.zip'
Content type 'application/zip' length 212201 bytes
(207 KB)
downloaded 207 KB
package 'exam' successfully unpacked and MD5 sums
checked
The downloaded binary packages are in
        C:\Users\Dell\AppData\Local\Temp\RtmpwTLuVC\
downloaded_packages
```

Next, run the command given below; also output is there:

```
> library(expm)
Loading required package: Matrix
Attaching package: 'expm'
The following object is mask from 'package:Matrix':
    Exam
```

Let's create a matrix in R using expm library.[7] Write the below lines of program code in the console of RStudio.

```
Matrix_1 <-matrix(1:25,nrow=5)
print(Matrix_1)
```

When you hit enter, the output will be like this:

```
> Matrix_1 <-matrix(1:25,nrow=5)
> print(Matrix_1)
      [,1] [,2] [,3] [,4] [,5]
[1,]    1    6   11   16   21
[2,]    2    7   12   17   22
[3,]    3    8   13   18   23
[4,]    4    9   14   19   24
[5,]    5   10   15   20   25
```

Another example:

```
> Matrix_2 <- matrix(sample(1:12,36,replace=TRUE),
nrow=6)
> Matrix_2
```

```
        [,1] [,2] [,3] [,4] [,5] [,6]
[1,]     12    1    4    9    8    6
[2,]      8    4    7    8    8    4
[3,]      5   10   10    9    7    4
[4,]      3   10    6   11    4    3
[5,]      7    4    6    3    7   12
[6,]      6    9    9    4    1   12
```

Power of matrix using two matrixes:

```
> Matrix_1 <-matrix(1:25,nrow=5)
> print(Matrix_1)
      [,1] [,2] [,3] [,4] [,5]
[1,]     1    6   11   16   21
[2,]     2    7   12   17   22
[3,]     3    8   13   18   23
[4,]     4    9   14   19   24
[5,]     5   10   15   20   25
> Matrix_2 <-matrix(1:25,nrow=5)
> print(Matrix_1)
      [,1] [,2] [,3] [,4] [,5]
[1,]     1    6   11   16   21
[2,]     2    7   12   17   22
[3,]     3    8   13   18   23
[4,]     4    9   14   19   24
[5,]     5   10   15   20   25
> Power_Matrix   <- Matrix_1%^%Matrix_2
> Power_Matrix
      [,1] [,2] [,3] [,4] [,5]
[1,]     1    6   11   16   21
[2,]     2    7   12   17   22
[3,]     3    8   13   18   23
[4,]     4    9   14   19   24
[5,]     5   10   15   20   25
```

Another example:

```
> Matrix<-matrix(sample(1:30,70,replace=TRUE),
ncol=10)
> Matrix
        [,1] [,2] [,3] [,4] [,5] [,6]
[1,]       3    9   29   12   20   13
[2,]      22   24   22    7   11    8
```

```
[3,]    15     7    19     8    26     3
[4,]    24    16    16     2    21     6
[5,]    30    21    15    16    10     7
[6,]    15    29    30    14    24     3
[7,]    19    25     3    17    12    10
       [,7]  [,8]  [,9] [,10]
[1,]    17    23     1    25
[2,]     9    18    29     2
[3,]     1    25    30    25
[4,]     6    21    23    14
[5,]    27     5    29     9
[6,]    25    28    26    24
[7,]    20    17    25    25
```

You can even multiple the numerics with matric like; first, create the matrix and then multiple it by 2:

```
> Matrix_2 <-matrix(1:25,nrow=5)
> print(Matrix_1)
      [,1] [,2] [,3] [,4] [,5]
[1,]     1    6   11   16   21
[2,]     2    7   12   17   22
[3,]     3    8   13   18   23
[4,]     4    9   14   19   24
[5,]     5   10   15   20   25
> Matrix_Result = Matrix_2 * 2
> print(Matrix_Result)
      [,1] [,2] [,3] [,4] [,5]
[1,]     2   12   22   32   42
[2,]     4   14   24   34   44
[3,]     6   16   26   36   46
[4,]     8   18   28   38   48
[5,]    10   20   30   40   50
```

Matrix is multiple with 3 in the following:

```
> Matrix<-matrix(sample(1:50,80,replace=TRUE),ncol=10)
> print(Matrix*3)
      [,1] [,2] [,3] [,4] [,5] [,6]
[1,]    21   42  132  147  129   12
[2,]    99   60    9   90  144   63
[3,]    33   51  102  132   18   45
```

```
[4,]    36    42     9   120    12   114
[5,]   147   150   117    42   114    90
[6,]   114    15   105     6    18    24
[7,]   126    60   123    39   102   114
[8,]   123    87    78    69    93    42
      [,7]  [,8]  [,9]  [,10]
[1,]   141   126    12    39
[2,]   105    51    87   120
[3,]    99   126    63    57
[4,]    57   132   135    12
[5,]   108    33   117   138
[6,]    99    27    78    69
[7,]   150   135    66   129
[8,]   129   102    21    18
```

This is the example of rnorm() with ncol.
Matrix<-matrix(rnorm(36,2,5),ncol=6)
print(Matrix)

```
> Matrix<-matrix(rnorm(36,2,5),ncol=6)
> print(Matrix)
           [,1]        [,2]       [,3]
[1,] -5.219795   9.204132  10.158959
[2,] -5.239433  14.355676   5.468737
[3,]  4.286827   8.712290   3.696239
[4,]  4.327034  -1.418014  -2.112402
[5,] -2.856190   5.778977   3.615762
[6,] -1.705022   5.845107   7.053443
           [,4]        [,5]        [,6]
[1,]  6.072929  9.9726634   3.1302871
[2,] 11.276494  0.7511379   1.1780292
[3,]  5.516676  8.4065138   7.1164443
[4,]  1.401199  9.2153035  -4.0005441
[5,]  3.679416  3.8946423   3.8451629
[6,]  1.651985  3.5372873   0.9661401
```

This is the example of rpois() with ncol.

```
> Matrix<-matrix(rpois(36,5),ncol=6)
> print(Matrix)
      [,1]  [,2]  [,3]  [,4]  [,5]  [,6]
[1,]     3     7     6     5     3     4
```

[2,]	6	4	3	6	4	6
[3,]	5	3	7	6	3	4
[4,]	3	9	4	5	5	6
[5,]	4	4	3	4	4	8
[6,]	8	6	2	2	6	7

Determinant of a Matrix in R

The determinant of the matrix AA, generally denoted |A||A|, is a scalar value that encodes some property of the matrix. In R you can use the det function to calculate it. The det() function in R is used to calculate the determinant of a given matrix.

Syntax:

```
det(x, …)
```

Parameters are x: matrix

Example:

```
> x <- matrix(c(3, 2, 6, -1, 7, 3, 2, 6, -1),
3, 3)
> print(x)
     [,1] [,2] [,3]
[1,]    3   -1    2
[2,]    2    7    6
[3,]    6    3   -1
> det(x)
[1] -185
```

Another example using transpose:

```
> # The R program to illustrate det function
> # First Initializing a matrix with
> x <- matrix(c(3, 2, 6, -1, 7, 3, 2, 6, -1), 3, 3)
> # 3 rows and 3 columns
> # Getting the matrix representation
> print(x)
     [,1] [,2] [,3]
[1,]    3   -1    2
[2,]    2    7    6
[3,]    6    3   -1
```

```
>
> # Getting the transpose of the matrix x
> y <- t(x)
> y
     [,1] [,2] [,3]
[1,]    3    2    6
[2,]   -1    7    3
[3,]    2    6   -1
>
> # Calling the det() function
> det(y)
[1] -185
```

Inverse of a Matrix in R

The inverse of a matrix is just the inverse of a matrix, as we do in normal arithmetic for a single number, which is used to solve equations to find the value of unknown variables. The inverse matrix is that matrix which, when multiplied by the original matrix, gives the identity matrix.

Finding the inverse of a matrix is one of the so common tasks when working with linear algebraic expressions. We can find the inverse value only for those matrices that are square and whose determinant is nonzero.

You can use the solve function to calculate the inverse of a matrix in R.

Example:

```
> # In this R program we find the inverse of a
Matrix
> # here we create 3 different vectors
> # using combine method.
> a1 <- c(2, 2, 3)
> a2 <- c(0, 1, 1)
> a3 <- c(5, 2, 4)
> # bind the three vectors into a matrix
> # using rbind() which is basically
> # row-wise binding.
> A <- rbind(a1, a2, a3)
> # print the original matrix
> print(A)
     [,1] [,2] [,3]
a1      2    2    3
a2      0    1    1
a3      5    2    4
```

```
> # Use the solve() function
> # to calculate the inverse.
> T1 <- solve(A)
> # print the inverse of the matrix.
> print(T1)
      a1 a2 a3
[1,] -2  2  1
[2,] -5  7  2
[3,]  5 -6 -2
```

There are two ways to find the inverse of a matrix:

- Using the solve() function: solve() is a general built-in function that is useful for solving the following linear algebraic equation.

- Using the inv() function: The inv() function is a built-in function that is mainly used to find the inverse of a matrix.

Example:

```
> # Create 3 different vectors.
> a1 <- c(2, 2, 2)
> a2 <- c(1, 1, 1)
> a3 <- c(1, 2, 1)
> # Bind the 3 matrices row-wise
> # using the rbind() function.
> A <- rbind(a1, a2, a3)
> # determinant of matrix
> print(det(A))
[1] 0
```

Rank of a Matrix in R

The rank of a matrix is a maximum number of columns (rows) that are linearly independent. There is no function to calculate the rank of a matrix, but we can make use of the QR function.

Example:

```
>   We assign matrix here,
> Matrix_2 <-matrix(1:25,nrow=5)
> print(Matrix_1)
      [,1] [,2] [,3] [,4] [,5]
```

```
[1,]    1    6   11   16   21
[2,]    2    7   12   17   22
[3,]    3    8   13   18   23
[4,]    4    9   14   19   24
[5,]    5   10   15   20   25
> QR(Matrix_2)$rank
[1] 2
```

R RELATIONAL OPERATORS

Logical operations simulate element-by-element decision operations based on a specified operator between operands, which then evaluate to a true or false Boolean value. Any other nonzero integer value is considered a TRUE value, whether complex or real. Relational operators are those that determine the relationship between two operands given to them. The following are six relational operations that the R programming language supports. The output is Boolean for all relational operators in the R programming language.

Operator	Explanation	Usage
<	The value of operand less than the second operand (also known as less than)	a < b
>	The value of operand greater than the second operand (also known as greater than)	a > b
==	The value of first operand must equal to the second operand (also known as equal than)	a = = b
<=	The value of first operand less than or equal to the second operand (less than equal to)	a <= b
>=	The value of first operand greater than or equal to the second operand (greater than equal to)	a > = b
!=	The value of first operand not equal to the second operand (not equal to)	a!=b

Relational operators perform comparison operations between the corresponding elements of the operands. The relational operator returns a Boolean TRUE if the first operand satisfies the relation compared to the second. TRUE is always considered higher than FALSE.

Let's discuss each of them:

Less Than (<) and Greater Than (<)

It usually returns TRUE if the corresponding element of the value of the first operand is less than the (<) value of the second operand.[8] Otherwise, it returns FALSE.

In the case of numerical values, it is pretty straightforward. For example, 2 is less than 4, so 2 < 4 will evaluate to TRUE, while 2 greater than 4 (2 > 4) will evaluate to FALSE.

For strings character, R uses the alphabet to sort them. So, "RStudio" > "Python" would evaluate to TRUE since "R" comes after "P" in the alphabet, and R console considers "RStudio" greater.

Example:

```
> # < and >
> # Less (<) than greater (>)
> 33 > 55
[1] FALSE
> 33 < 55
[1] TRUE
> # Alphabetically less than (>) greater than (>)"
> "R programming " > "Python"
[1] TRUE
> " PHP " < " HTML"
[1] FALSE
```

Example:

```
> list_1 <- c(TRUE, 0.1," HTML ")
> list_2 <- c(0,0.1," PHP ")
> print(list_1<list_2)
[1] FALSE FALSE   TRUE
```

Greater Than (>)

It usually returns TRUE if the corresponding element of the first operand is greater than the second operand; otherwise, it returns FALSE.

We can also check if one R object is greater than or equal to or less than or equal to another R object. For this, we can use the less than sign or the greater than sign together with the equals sign.

Example:

```
> list_1 <- c(TRUE, 0.1," HTML ")
> list_2 <- c(0,0.1," PHP ")
> print(list_1 > list_2)
[1]   TRUE FALSE FALSE
```

Less Than Equal (<=)

It usually returns TRUE if the corresponding element of the first operand is less than or equal to the element of the second operand, otherwise, it returns FALSE.

Example:

```
> list_1 <- c(TRUE, 0.1,"HTML")
> list_2 <- c(0,0.1,"PHP")
> print(list_1 <= list_2)
[1] FALSE   TRUE   TRUE
```

Greater Than Equal (>=)

It usually returns TRUE if the corresponding element of the first operand is greater than or equal to the element of the second operand, otherwise, it returns FALSE.

Example:

```
> # <=   and   >=
> 33 >= 33
[1] TRUE
> 33 >= 55
[1] FALSE
> 33 <= 55
[1] TRUE
> 55 <= 55
[1] TRUE

> # Another example,
> list_1 <- c(TRUE, 0.1,"HTML")
> list_2 <- c(0,0.1,"PHP")
> print(list_1 >= list_2)
[1]   TRUE   TRUE FALSE

> # Something Extra
> "PHP" < "HTML"
[1] FALSE
> TRUE <= FALSE
[1] FALSE
> (1+2) > 4
[1] FALSE
```

```
> # Comparison of numeric
> (-6 * 1 + 2) >= (-12 + 1)
[1] TRUE
> # Comparison of character strings
> "R " <= "R programming "
[1] TRUE
> # Comparison of logical
> TRUE > FALSE
[1] TRUE
```

Instead of doing one-by-one comparison of value, you can combine list using c ():

```
> user_1_marks <- c(50,80,80,60)
> print(user_1_marks)
[1] 50 80 80 60
> user_2_marks <- c(60, 40, 60, 80)
> print(user_2_marks)
[1] 60 40 60 80
> user_1_marks < user_2_marks
[1]   TRUE FALSE FALSE   TRUE
```

Not Equal to (!=)

The unequal operator in R is one of the relational operators and is the opposite of the equality operator. The not equals operator is written as an exclamation point followed by an equals sign (!=).

It returns TRUE if the value of the first operand is not equal to the value of the second operand, otherwise, it returns FALSE.

Example:

```
> list_1 <- c(TRUE, 0.1,"PHP")
>          list2 <- c(0,0.1,"Python")
>          print(list_1!=list_2)
[1]   TRUE FALSE FALSE

> a <- "R"
> b <- "R"
> print(a!=b)
[1] FALSE
```

Program to understand all the relational operators:

```
> # Whole program to understand
> # the use of Relational operators
> vec1 <- c(1, 2)
> vec2 <- c(4, 3)
> # performing operations on Operands
> cat(" First value less than Second value :",
vec1<vec2, "\n")
 First value less than Second value : TRUE TRUE
> cat("First value less than equal to Second value :",
vec1<=vec2, "\n")
First value less than equal to Second value : TRUE
TRUE
> cat ("First value greater than Second value :",
vec1>vec2, "\n")
First value greater than Second value : FALSE FALSE
> cat("First value greater than equal to Second value :",
vec1>=vec2, "\n")
First value greater than equal to Second value : FALSE
FALSE
> cat("First value not equal to Second value :",
vec1!=vec2, "\n")
First value not equal to Second value : TRUE TRUE
```

Assignment Operators

The assignment operator is used to assign a new value to a variable, property, event, or indexer element in the C# programming language.[9] These assignment operators can also be used for logical operations such as bitwise logical operations or operations with integral operands and Boolean operands.

It allows the assignment operator to be used with these types. Assignment operators are used to assign values to different data objects in R. Objects can be integers, vectors, or functions. These values are then stored under the assigned variable names. There are two types of assignment operators such as Left and Right.

These include:

- +=

- −=

- $*=$
- $/=$
- $\%=$
- $\&=$
- $|=$
- $\wedge=$
- $<<=$ and $>>=$

The <- and = operators can be used almost interchangeably to assign a variable in the same environment. The <<- operator is used for assignments to variables in parent environments (rather global assignments). True assignments, even when available, are rarely used.

Left assignment (<- or <<- or =): It assigns a value to a vector.

Example:

```
> val_1 = c("ABC", TRUE)
> print (val)
[1] "ABC"   "TRUE"
> val_2 <<- c("ABC", TRUE)
> print (val)
[1] "ABC"   "TRUE"

> val_2 <- c("abc", TRUE)
> print (val)
[1] "abc"   "TRUE
```

Right assignment (-> or ->>): It assigns a value to a vector.

Example:

```
> # here we assign value to val_2,
> val1 <- c("ABC", TRUE)
> print (val1)
[1] "ABC"   "TRUE"
# here we assign -> using this operator,
> c("ABC", TRUE) -> val2
> print(val2)
[1] "ABC"   "TRUE"
```

R LOGICAL OPERATORS

Logical operators in the R programming language work only for the logical, numeric, and complex basic data types and vectors of these basic data types.[10] R uses the AND, OR, and NOT operators.

- Logical operators

- AND operator &

- OR operator |

- NOT an operator !

Operator	Description	Usage
&	a & b	It is called element-wise logical operator AND. It combines each element of the first vector with the corresponding element of the second vector and outputs TRUE if both elements are TRUE
|	a | b	It is called element-wise logical operator AND. It combines each element of the first vector with the corresponding element of the second vector and outputs TRUE if both elements are TRUE
!	!a	This is called the logical NOT operator. It takes every element of the vector and gives the opposite Boolean value
&&	a && b	It is called the logical operator AND. Takes the first element of both vectors and returns TRUE only if both are TRUE
||	a || b	It is called the logical OR operator. Takes the first element from both vectors and returns TRUE if either is TRUE

AND Operator "&"

The AND operator takes two logical values and returns TRUE only if both values are TRUE that means TRUE & TRUE evaluates to TRUE, but that FALSE & TRUE, TRUE & FALSE, and FALSE & FALSE evaluate to FALSE; the reason is that when one FALSE occurs, that whole answer is FALSE.

Example:

```
> TRUE & TRUE
[1] TRUE
> FALSE & FALSE
[1] FALSE
> TRUE & FALSE
[1] FALSE
> FALSE & TRUE
[1] FALSE
```

Instead of using Boolean values, we can use comparison results. You can suppose we have a variable x equal to 7. To check if the variable is greater than 4 but less than 20, we can use x greater than 4 and x less than 20. For example,

```
x <- 7
x > 4 & x < 20
```

The first part x > 4 will evaluate to TRUE because 7 is greater than 4. The second part x < 20 will also evaluate to TRUE because 7 is also less than 20. So the result of the expression is TRUE because TRUE & TRUE is TRUE. This makes sense because 7 lies between 4 and 20.

R Miscellaneous Operators

These operators do not fall into any of the categories and above but are significantly important during R programming for manipulating data.

Operator	Explanation	Usage
:	It is a colon operator. It creates a series of numbers from left operand to the right operand	a:b
%in%	It identifies if an element (a) belongs to a vector (b)	a %in% b
%*%	It performs multiplication of a vector with its transpose	A %*% t(A)

Example:

```
> a = 25:21
> print ( a )
[1] 25 24 23 22 21

> a = c(15, 17, 16)
> b = 27
> print ( b %in% a )
[1] FALSE

> M = matrix(c(1,2,3,4), 2, 2, TRUE)
> print ( M %*% t(M) )
     [,1] [,2]
[1,]    5   11
[2,]   11   25
```

CHAPTER SUMMARY

In this chapter, we learned about the basic use of variable, types of it, and where and how to use it. It also covered some examples of valid and invalid variable used in R. Another topic includes operators and its types.

NOTES

1. Variable in R – https://www.datamentor.io/r-programming/variable-constant/, accessed on September 28, 2022.
2. Constants in R – https://www.w3schools.in/r-programming/variables-constants-vectors/, accessed on September 28, 2022.
3. Constants in R – https://www.datamentor.io/r-programming/variable-constant/, accessed on September 28, 2022.
4. Operators in R – https://www.tutorialkart.com/r-tutorial/r-operators/, accessed on September 28, 2022.
5. Matrix Operations – https://r-coder.com/matrix-operations-r/#Addition_and_substraction, accessed on September 29, 2022.
6. Matrix in R – https://www.tutorialkart.com/r-tutorial/r-matrix-multiplication/, accessed on September 29, 2022.
7. Create Matrix Using Library – https://www.tutorialspoint.com/how-to-find-power-of-a-matrix-in-r, accessed on September 29, 2022.
8. Relational Operator – https://towardsdatascience.com/the-ultimate-guide-to-relational-operators-in-r-6d8489d9d947, accessed on September 30, 2022.
9. Assignment Operator – https://www.techopedia.com/definition/25583/assignment-operator-c, accessed on September 30, 2022.
10. Logical Operator – https://towardsdatascience.com/the-complete-guide-to-logical-operators-in-r-9eacb5fd9abd, accessed on September 30, 2022.

## CHAPTER 4

Loops and Decision-Making in R

IN THIS CHAPTER

➢ Introduction

➢ Control Structures in R Programming

The previous chapter is about valid data with data types, now we will use that variable with various decision-making statement and looping in R. Today we will study the structures and commands that control the flow of R program.

In R programming, as in other languages, there are several cases where you may wish to conditionally execute any code. R's "if" and "switch" functions can be implemented here if you have already programmed condition-based code in other languages. Vectorized conditional implementation through the ifelse() function is also a characteristic of R. In this chapter, you will look at all these conditional statements that R provides for programmers to write.

Decision-making is an important feature of any programming language. It allows to change the order of code execution based on certain conditions. These structures can use in all combinations with each other for different working scenarios and implementations. The way to learn more about their combinations and results is to practice.

DOI: 10.1201/9781003358480-4

CONTROL STRUCTURES IN R PROGRAMMING

R provides various standard control structures for the requirements. An expr expression consists of multiple statements that can be enclosed in braces {}. Whenever possible, it is more efficient to use built-in functions in R rather than control structures. It facilitates the flow of execution to be controlled inside the function. Control structures define program flow. A decision is then made after evaluating the variable.

There are many situations where you don't just want to execute one command after another: you actually need to control the flow of execution as well that means that you just want to run some code if the condition is met.[1] In that case, the control flow statements are implemented in the R program. There is the common structure of how control flow can be handled using conditional statements in R programming. Decision-making statements are also referred to as selection statements.

Following is the table of control statement in R.

Statement	Description	Syntax

If statements	A block of statements executed only if the specified test expression is true	`if(boolean _ expression) {`
		`    // If the Boolean statement is true, then statement(s) will be executed.`
		`}`

If–else statements	When we want to execute some block of code if a condition is true and then another block of code if the condition is false, in that case, we use the if … else statement	`if (condition1) {`
		`    expr1`
		`    } else if (condition2) {`
		`}`

If–else–if statements	When we want to add more condition checks in a single if–else statement, then we can easily add more conditions using the swift if–else–if else statement. In the if–else–if else statement, we have the option to add alternative else–if statements, but we are limited to only one if and else block in the statement	`if (condition1) {`
		`    expr1`
		`    } else if (condition2) {`
		`    expr2`
		`    } else if (condition3) {`
		`    expr3`
		`    } else {`
		`    expr4`
		`}`

(Continued)

Statement	Description	Syntax
The switch statement	Evaluates an expression against multiple cases to identify the block of code to be executed	`Switch(expression, case _ 1, case _ 2, ..........)`

For loops	It repeats through sequences to perform repeated tasks. They work with an iterable variable to go through a sequence	`for(value in vector){`
		`    statements`
		`    ....`
		`    ....`
		`}`

A while loop	R evaluates a condition. If the condition results to TRUE, it goes through the code block, if the condition evaluates to FALSE, it exits the loop. A while loop in R continues to loop through a closed block of code until the condition is TRUE	`while(expression){`
		`    statement`
		`    ....`
		`    ....`
		`}`

Repeat statement	It uses inside a loop (repeat, for, while) to stop iterations and control the flow outside the loop	`repeat {`
		`    statements`
		`    ....`
		`    ....`
		`    if(expression) {`
		`      break`
		`    }`
		`}`

Break statement	It is used to repeat a block of code over and over. In a repeat loop, there is no condition check to exit the loop	`Break`
Next statement	Next statement in R It actually jumps to evaluate the condition holding the current loop. The next command allows skipping the current iteration of the loop without terminating it	`next`
Return statement in R	It requires some functions to do processing and return back the result	`Return (expression)`

Now we will discuss various control structures in R one by one in brief.

If Statement in R

The simplest form of decision controlling the statement for conditional execution is the "if" statement.[2] The "if" gives you a logical value (more exactly, a logical vector having length 1) and that carries out the next statement when that value becomes TRUE. In other words, an "if" statement is having a Boolean expression followed by single or multiple statements. The block of code inside if statement will only be executed if the Boolean expression evaluates to true. If the statement results to false, the code given after the condition is executed.

This statement consists of Boolean expressions followed by one or more statements. R if statement is the simplest decision statement that helps us make a decision based on a condition. R if statement in R is a conditional programming statement that executes a function and displays information if it proves to be true.

Syntax:

```
if(boolean_expression) {
     // here you can write your code the statement
only execute when your expression is true.
}
```

If the Boolean expression evaluates to be true, then a block of code inside the if statement will be executed. If Boolean expression results to be false, then the set of code after end of the if statement (after closing curly brace) will be executed.

Let's have a look at few examples of if statement.

First example:

```
> values <- 1:10
> if (sample(values,1) <= 10)
+       print(paste("Value is ", values))
  [1] "Value is  1"   "Value is  2"
  [3] "Value is  3"   "Value is  4"
  [5] "Value is  5"   "Value is  6"
  [7] "Value is  7"   "Value is  8"
  [9] "Value is  9"   "Value is  10"
```

Second example:

```
> a <- 13
> b <- 20
>
> if (b > a) {
```

```
+        print("b is greater than a")
+ }
[1] "b is greater than a"
```

Third example:

```
> x <- 18
> y <- 26
> count = 0
> if (x < y)
+ {
+        cat (x, "is a smaller number\n")
+        count=1
+ }
18 is a smaller number
> if ( count == 1 ){
+        cat ("The block is successfully execute")
+ }
The block is successfully execute
```

Fourth example:

```
> x <- 1
> y <- 24
> count = 0
> while ( x < y ){
+        cat (x, "\n")
+        x = x + 2
+        if (x == 15)
+              break
+ }
1
3
5
7
9
11
13
```

NESTED IF STATEMENTS

You can have if statements inside if statements, it is called nested if statements.

Example:

```
> x <- 24
> if (x < 18) {
+      print(" Age is under 18 ")
+    · if (x > 20) {
+          print("and above 10!")
+      } else {
+          print("but not above 10.")
+      }
+ } else {
+      print("  Age is above 18. ")
+ }
[1] "  Age is above 18. "
```

If–Else Statement

In an if statement, the inner code is executed when the condition is true.[3] The code outside the if block will be executed when the if condition is false.

There is another type of decision statement known as the if–else statement. An if–else is an if statement followed by an else statement. In if–else statement, the else part will be executed when the Boolean expression is false. If Boolean expression evaluates to true, then if block will be executed or the else block will be executed.

Syntax:

```
if ( expression ) {
# here is the statement of if code
} else {
# here is the statement of else code
}
```

First example:

```
> # Here we assign variable a
> a <- 10
> #checking Boolean condition
> if( a < 20){
+      # if the condition is true then print the
following
+      cat(" Value of a is less than 20\n")
+ }else{
```

```
+      # if the condition is false then print the
following
+      cat(" Value of b is not less than 20\n")
+ }
Value of a is less than 20
> cat("The value of a is", a)
The value of a is 10
```

Second example:

```
> x <- c(" We "," our ", " learning "," R ","
Programming ","concepts")
>
> if(" R " %in% x) {
+      print("key is found")
+ } else {
+      print("key is not found")
+ }
[1] "key is found"
```

Nested If–Else Statements

You can also have if–else statements inside if–else statements; this is called nested if–else statements.

Example:

```
> x <- 24
>
> if (x < 18) {
+      print(" Age is under 18 ")
+      if (x > 20) {
+           print("and above 10!")
+      } else {
+           print("but not above 10.")
+      }
+ } else {
+      print(" Age is above 18. ")
+ }
[1] " Age is above 18. "
```

R If–Else–If Statement

In R, the if–else–if statement allows to add an alternative set of test conditions in an if–else statement using an else–if statement and a single else for the if condition. The if–else–if statement is used to select one of several blocks of code to be executed.

Syntax:

```
if (condition1) {
  // here will be the code of if
}else if (condition2) {
  // here will be the code of else
}
.

.

else if (conditionN) {
  // here will be the code of else
}else {
  // here will be the code of else
}
```

Example:

```
> a <- 8
> b <- 10
> print (" R If..else..if Statement ")
[1] " R If..else..if Statement "
> if (a < b) {
+      print ("a is smaller than b")
+ } else if(a == b){
+      print ("a and b are equal")
+ } else{
+      print ("b is greater than a")
+ }
[1] "a is smaller than b"
```

R Switch Statement

In R, the switch statement evaluates the value of a variable/expression for equality against multiple case values to identify the block of code to be executed.

Syntax:

```
switch(expression, case1, case2, case3,....,
caseN)
```

The following rules apply to the switch statement:

- If the expression value is not a character string.

- You can have any various numbers of case statements in a switch.

- If there is more than one match, the first matching element is returned.

- There is no default value/case.

- If no match is found and an unnamed element exists, its value is returned.

First example:

```
> day <- 2
> week <- switch(
+       day,
+       "Sunday",
+       "Monday",
+       "Tuesday",
+       "Wednesday",
+       "Thursday",
+       "Friday",
+       "Saturday",
+
+ )
> print(" R Switch Statement")
[1] " R Switch Statement"
> print(week)
[1] "Monday"
```

Second example:

```
> y = "18"
> x = switch(
+       y,
+       "9"=" Sam ",
+       "12"=" Lavi ",
+       "18"=" Jazz ",
+       "21"= " Som "
+ )
```

```
>
> print (x)
[1] " Jasz "
> y = "12"
> x = switch(
+      y,
+      "9"=" Sam ",
+      "12"=" Lavi ",
+      "18"=" Jasz ",
+      "21"= " Som "
+ )
>
> print (x)
[1] " Lavi "
```

R Repeat Statement

A loop repeatedly executes a block of code until it meets an explicit condition to break and exit the loop. The loop iteration construct has no conditional control over iterations, so it is mandatory to define an explicit condition with a break statement inside the loop body, which allows the program to exit the loop. Otherwise, it will give an infinite loop.

Syntax:

```
repeat {
//statements
}
```

Example:

```
> ctr <- 1
> repeat {
+      print("R, Programming Language")
+      ctr = ctr+1
+      if (ctr >= 4){
+          break
+      }
+ }
[1] "R, Programming Language"
[1] "R, Programming Language"
[1] "R, Programming Language"
```

Next Statement

In R, the next statement allows you to skip the current iteration of any loop and continue with the next iteration. When the next statement is found in the loop, the rest of the statements skips in the loop body for the current iteration and return program execution to the very first statement in the loop body. It does not terminate the loop but continues with the next iteration.

Syntax:

```
if (condition) {
next
}
```

Example:

```
> print("R Next Statement")
[1] "R Next Statement"
> count <- 1
> while (count <= 10) {
+       count = count + 1
+       if (count == 6) {
+               print("6 is skipped")
+               next
+       }
+       print("Inside loop")
+ }
[1] "Inside loop"
[1] "Inside loop"
[1] "Inside loop"
[1] "Inside loop"
[1] "6 is skipped"
[1] "Inside loop"
[1] "Inside loop"
[1] "Inside loop"
[1] "Inside loop"
[1] "Inside loop"
```

Return Statement

We will require some functions to do processing and return back the result. It gets with the return() statement in R. Following is the syntax, return(expression).

Syntax:

return(expression)

Example:

```
> check <- function(x) {
+       if (x < 0) {
+            result <- "Negative"
+       } else if (x > 0) {
+            result <- "Positive"
+       } else {
+            result <- "Zero"
+       }
+       return(result)
+ }
> check(-10)
[1] "Negative"
> check(10)
[1] "Positive"
> check(0)
[1] "Zero"
```

Break Statement

It allows you to interrupt or terminate the execution of the loop and pass execution to the next statement that follows the loop. It is almost always used with an if–else construct. When the break statement is used inside a nested loop, it allows the break and termination of the innermost loop in which it is used.

First example:

```
> print("R Break Statement")
[1] "R Break Statement"
> count <- 0
> while (count <= 10) {
+       count = count + 1
+       if (count == 5) {
+            break
+       }
+       print("Inside loop")
+ }
[1] "Inside loop"
[1] "Inside loop"
[1] "Inside loop"
[1] "Inside loop"
> print("Out of while loop")
[1] "Out of while loop"
```

Explanation: Initially, the variable count is 0. Then a while loop is executed until the count is less than 10. The variable count is incremented by 1 (count = count + 1) at each iteration one by one. Next, we have statement that checks if the number of variables is 5; if it returns TRUE, it will only cause the loop to break or exit. Then print() statement will be executed on each iteration of the while loop until the loop is terminated. After that, there is the final print() statement outside the while loop.

Second example:

```
> x = 1: 5
> for (i in x) {
+       if (i == 2) {
+           next
+       }
+       print(i)
+ }
[1] 1
[1] 3
[1] 4
[1] 5
```

LOOPING IN R

The for loop is the control flow statement.[4] A for loop is used to iterate over a vector. It is similar to a while loop. There is only one difference between for and while, that is, in the while loop, the condition is checked before the body is executed, but the condition in the for loop is checked after the execution.

Syntax:

```
for (initialization ; expression; statement)
{
    // statements inside the body of the loop
}
```

Example:

```
> # Here we Create language vector
> language <- c('PHP', 'R', "Python", 'Ruby',
'C++','C')
```

```
> # Create the for statement
> for ( i in language){
+       print(i)
+ }
[1] "PHP"
[1] "R"
[1] "Python"
[1] "Ruby"
[1] "C++"
[1] "C"
```

R ifelse() Function

Vectors are the basic building block of R programming. Most functions in R take a vector as input and output a resulting vector. This vectorization of the code will be much faster than applying the same function to each element of the vector individually. Similarly, there is a vector equivalent form of the if–else statement in R, the ifelse() function.

Test_expression here must be a Boolean vector (or an object that can be coerced to Boolean). The return value is a vector with the same length as expression. The returned vector has an element of x if the corresponding value of expression is TRUE or of y if the corresponding value of expression is FALSE.

CHAPTER SUMMARY

We studied about various looping statements like for, while, and do while with some more about break, next, and return, and also other decision-making loops like if, else, else–if, and so on.

NOTES

1. Flow Control – https://www.w3schools.in/r-programming/conditional-statements, accessed on October 1, 2022.
2. Flow Control Statement – https://www.w3schools.in/r-programming/con-ditional-statements, accessed on October 1, 2022.
3. If–Else – https://techvidvan.com/tutorials/r-control-structures/, accessed on October 1, 2022.
4. Looping in R – https://www.javatpoint.com/r-for-loop, accessed on October 1, 2022.

## CHAPTER 5

Functions and Strings

IN THIS CHAPTER

- ➤ Introduction
- ➤ Arguments
- ➤ Function Types
- ➤ Global Operator
- ➤ Built-in Function
- ➤ and more

In the last chapter, we have gone through all the decision-making statement and loops in the R programming language. It is the commonly used concept in every programming language. Now we have another chapter function and string. You will get deep knowledge of using function and string here.

INTRODUCTION

All the variables we use in the program must be stored somewhere. This somewhere can be called an environment in R.[1] It is closely related to lists in which they are used to store different types of variables together. There are two situations where you can encounter the environment. Whenever a function is called, all variables described by the function are stored in the environment belonging to that specific function. When you load a package, the functions in that package are stored in the environment on the search path.

Defining a Function

You can write your functions to perform repeated operations with a single command. Let's start by defining your "my_function" function and the input parameters that the user will put into the function. You then define the operation you wish to program in the body of the function in braces ({}). Finally, you must assign the result (or output) of your function in the return statement.

R function is created by using the keyword function.

Syntax:

```
function_name <- function(arg_1, arg_2, ...) {
    # here you can write function body
}
```

Above is the general syntax of an R function.[2] In the following:

- function: It is a keyword which is used to define a function.

- func_name: It replaces with the name of the function.

- arg_1, arg_2, ...: It represents the list of the parameters need to be passed when a function call is made.

Functional Components[3]

The different parts of the function are as follows:

- Function name: It is the actual name of the function and stored in the R environment as an object with this name.

- Arguments: An argument is a placeholder. When the function is invoked, you can pass the value of the argument. Arguments are optional, that is, the function must not contain any arguments. Arguments can also have default values.

- Function body: The body of the function contains a collection of statements that define what the function does.

- Return value: The value of a function is the last expression in the function body to be evaluated.

A function gives you a way to wrap a set of commands, meaning to perform any particular task, and name it so that it can be called later from anywhere in the program. Functions make it easy to divide the entire program into subunits that perform a specific task for the program, thus improving the modular approach and increasing the reusability of the program code. In calling a function, we pass information as its parameter, and the function can either return some value to the point from which it was called or return nothing.

Now let's look at this process with an example. We will define a Fahrenheit to Celsius function that converts temperatures from Fahrenheit to degree Celsius as given below.

```
> f_to_c <- function(temp_in_F) {
+      temp_in_C <- (temp_in_F - 32) * 5 / 9
+      return(temp_in_C)
+ }
> f_to_c(10)
[1] -12.22222
> f_to_c(5)
[1] -15
```

Arguments

Information can be passed to functions as arguments.[4] Arguments are listed after the function name in parentheses. You can add many arguments as you want, only separate them with a comma.

The example has a function with one argument (fname). When function is called, we pass the first name, which is used inside the function to print the full name.

```
> new_function <- function(fname) {
+    paste(fname, "R Programming")
+ }
> new_function(" Python ")
[1] " Python  R Programming"
> new_function(" PHP ")
[1] " PHP  R Programming"
> new_function(" Ruby ")
[1] " Ruby  R Programming"
```

Number of Arguments

By default, a function should be called with the right number of arguments; meaning that if function expects two arguments, you have to call the function with two arguments, not more and not less.

Example:

The following function takes two arguments and gets two arguments (parameter).

```
> new_function <- function(first_name, last_name) {
+     paste(first_name, last_name)
+ }
>
> new_function(" Sam ", " Noch ")
[1] " Sam   Noch "
```

Default Parameter Value

The following example shows the use of default parameter value. If we call function without an argument, it uses the default value.

```
> new_function <- function(country = "India") {
+     paste("I live in", country)
+ }
>
> new_function(" UK ")
[1] "I live in  UK "
> new_function(" USA ")
[1] "I live in  USA "
> new_function() # will get the default value, which
is India
[1] "I live in India"
> new_function(" UAE ")
[1] "I live in  UAE "
```

FUNCTION TYPES

There are mainly three types of functions in R programming:[5]

- Primitive function

- Infix function

- Replacement function

Let's discuss each of them.

Primitive Function

In general, a function consists of three parts:

- Formals(), a list of arguments that control how the function is called.

- Body(), the code inside the function.

- Environment() is a data structure that specifies how the function finds values associated with names.

The formals are defined explicitly whenever you create a function, but the environment is specified implicitly based on where the user defines the function. But there is an exception to the rule that a function has three components, functions call code directly. These functions are known as primitive functions. The primitive functions exist primarily in C, not R, so their formals(), body(), and environment() are NULL. These features are only found in the basic package. Primitive functions are not easy to write but are highly efficient. They are of two types, either built-in or special.

Following is the complete list of the primitive function in R:

```
> names(methods:::.BasicFunsList)
```

- [1] "$"

- [2] "$<-"

- [3] "["

- [4] "[<-"

- [5] "[["

- [6] "[[<-"

- [7] "%*%"

- [8] "xtfrm"

- [9] "c"

- [10] "all"

- [11] "any"

- [12] "sum"

- [13] "prod"
- [14] "max"
- [15] "min"
- [16] "range"
- [17] "is.matrix"
- [18] ">="
- [19] "cosh"
- [20] "cummax"
- [21] "dimnames<-"
- [22] "as.raw"
- [23] "log2"
- [24] "tan"
- [25] "dim"
- [26] "as.logical"
- [27] "^"
- [28] "is.finite"
- [29] "sinh"
- [30] "log10"
- [31] "as.numeric"
- [32] "dim<-"
- [33] "is.array"
- [34] "tanpi"
- [35] "gamma"
- [36] "atan"
- [37] "as.integer"
- [38] "Arg"
- [39] "signif"

- [40] "cumprod"
- [41] "cos"
- [42] "length"
- [43] "!="
- [44] "digamma"
- [45] "exp"
- [46] "floor"
- [47] "acos"
- [48] "seq.int"
- [49] "abs"
- [50] "length<-"
- [51] "sqrt"
- [52] "!"
- [53] "acosh"
- [54] "is.nan"
- [55] "Re"
- [56] "tanh"
- [57] "names"
- [58] "cospi"
- [59] "&"
- [60] "anyNA"
- [61] "trunc"
- [62] "cummin"
- [63] "levels<-"
- [64] "*"
- [65] "Mod"
- [66] "|"

- [67] "names<-"
- [68] "+"
- [69] "log"
- [70] "lgamma"
- [71] "as.complex"
- [72] "asinh"
- [73] "-"
- [74] "sin"
- [75] "/"
- [76] "as.environment"
- [77] "<="
- [78] "as.double"
- [79] "is.infinite"
- [80] "is.numeric"
- [81] "rep"
- [82] "round"
- [83] "sinpi"
- [84] "dimnames"
- [85] "asin"
- [86] "as.character"
- [87] "%/%"
- [88] "is.na"
- [89] "<"
- [90] ">"
- [91] "I'm"
- [92] "%%"
- [93] "trigamma"

- [94] "=="
- [95] "cumsum"
- [96] "atanh"
- [97] "sign"
- [98] "ceiling"
- [99] "Conj"
- [100] "as.call"
- [101] "log1p"
- [102] "expm1"
- [103] "("
- [104] ":"
- [105] "="
- [106] "@"
- [107] "{"
- [108] "~"
- [109] "&&"
- [110] ".C"
- [111] "baseenv"
- [112] "quote"
- [113] "::"
- [114] "<-"
- [115] "is.name"
- [116] "if"
- [117] "||"
- [118] "attr<-"
- [119] "untracemem"
- [120] ".cache_class"

- [121] "substitute"
- [122] "interactive"
- [123] "is.call"
- [124] "switch"
- [125] "function"
- [126] "is.single"
- [127] "is.null"
- [128] "is.language"
- [129] "is.pairlist"
- [130] ".External.graphics"
- [131] "globalenv"
- [132] "class<-"
- [133] ".Primitive"
- [134] "is.logical"
- [135] "enc2utf8"
- [136] "UseMethod"
- [137] ".subset"
- [138] "proc.time"
- [139] "enc2native"
- [140] "repeat"
- [141] ":::"
- [142] "<<-"
- [143] "@<-"
- [144] "missing"
- [145] "nargs"
- [146] "isS4"
- [147] ".isMethodsDispatchOn"

- [148] "forceAndCall"

- [149] ".primTrace"

- [150] "storage.mode<-"

- [151] ".Call"

- [152] "unclass"

- [153] "GC.time"

- [154] ".subset2"

- [155] "environment<-"

- [156] "emptyenv"

- [157] "seq_len"

- [158] ".External2"

- [159] "is.symbol"

- [160] "class"

- [161] "on.exit"

- [162] "is.raw"

- [163] "for"

- [164] "is.complex"

- [165] "list"

- [166] "invisible"

- [167] "is.character"

- [168] "oldClass<-"

- [169] "is.environment"

- [170] "attributes"

- [171] "break"

- [172] "return"

- [173] "attr"

- [174] "tracemem"

- [175] "Next"

- [176] ".Call.graphics"

- [177] "standardGeneric"

- [178] "is.atomic"

- [179] "retracement"

- [180] "expression"

- [181] "is.expression"

- [182] "call"

- [183] "is.object"

- [184] "pos.to.env"

- [185] "attributes<-"

- [186] ".primUntrace"

- [187] "...length"

- [188] ".External"

- [189] "oldClass"

- [190] ".Internal"

- [191] ".Fortran"

- [192] "browser"

- [193] "is.double"

- [194] ".class2"

- [195] "while"

- [196] "nzchar"

- [197] "is.list"

- [198] "lazyLoadDBfetch"

- [199] "...elt"

- [200] "...names"

- [201] "is.integer"

- [202] "is.function"

- [203] "is.recursive"

- [204] "seq_along"

- [205] "unlist"

- [206] "as.vector"

- [207] "lengths"

Here you can check whether a function is a primitive function or not in R programming using is.primitive()

Define: is.primitive() function is used to check if a function is a primitive function that is either a built-in function or a special function.

Syntax:

```
is.primitive(func).
```

Example:

```
> # The program to understand
> # the use of is.primitive function
> # here calling is.primitive() function
> is.primitive(1)
[1] FALSE
> is.primitive(is.primitive)
[1] FALSE
> is.primitive(sum)
[1] TRUE
> is.primitive(prod)
[1] TRUE
> is.primitive(tan)
[1] TRUE
> is.primitive(range)
[1] TRUE
> is.primitive(min)
[1] TRUE
> is.primitive(max)
[1] TRUE
```

Infix Functions

In R, most functions are "prefixed," meaning the function name comes before the arguments, which are enclosed in parentheses: fun(a,b).[6] For infix functions, the name is between the arguments and fun b. For example, operators like + and – are actually infix functions; these operators call a background function. a+b is actually converted to its infix equivalent '+'(a, b).

Example:

```
>   2 + 3
[1] 5
>   '+'(2,3)
[1] 5
```

INFIX OPERATORS IN R

The following is a list of predefined infix operators available in R.

- %%: It is remainder operator
- %/%: It is integer division
- %*%: It is matrix multiplication
- %o%: It is outer product
- %x%: It is Kronecker product
- %in%: It is matching operator

Example:

```
> # R program to illustrate
> # Infix function
>
> '%Greater%' <- function(a, b)
+ {
+       if(a > b) print(a)
+       else if(b > a) print(b)
+       else print("equal")
+ }
> 5 %Greater% 8
[1] 8
> 2300 %Greater% 66
[1] 2300
```

USER-DEFINED INFIX OPERATOR

In R, a user-defined infix operator can be created as follows:

```
'%func_name%' <- function(arg1, arg2) {
  # function body
}
```

Example:

```
> '%pwr%' <- function(x,y)
+ {
+        return(x^y)
+ }
> res <- 5 %pwr%  2
> print(" R Infix Operator ")
[1] " R Infix Operator "
> print(res)
[1] 25
```

THE gR GLOBAL VARIABLES

The variables created outside of a function are known as global variables.[7] Global variables can be used by everyone, inside of functions and outside, for example, to create a variable outside of a function and use it inside the function.

```
> txt <- "You are learning R Programming"
> my_function <- function() {
+        paste("R is", txt)
+ }
>
> my_function()
[1] "R is You are learning R Programming"
```

THE GLOBAL ASSIGNMENT OPERATOR

Basically, when you create a variable inside a function, then that variable is local and can only be used inside that function. To create a variable globally inside a function, you can use the global assignment operator <<-

Example:

```
> my_function <- function() {
+        txt <<- "Programming Language"
+        paste("R is", txt)
+ }
```

```
> my_function()
[1] "R is Programming Language"
> print(txt)
[1] "Programming Language"
```

TYPES OF FUNCTIONS

Function Call

The function can be called with an argument, without an argument, and also with a default value.

Example:

```
> # create a function cube
> # without an argument
> cube <- function()
+ {
+      for(i in 1:10)
+      {
+           print(i^3)
+      }
+ }

> # calling function cube without an argument
> cube()
[1] 1
[1] 8
[1] 27
[1] 64
[1] 125
[1] 216
[1] 343
[1] 512
[1] 729
[1] 1000
```

Another example:

```
> # create a function factorial
> # with a numeric argument n
> factorial <- function(n)
+ {
+      if(n==0)
```

```
+       {
+               return(1)
+       }
+       else
+       {
+               return(n * factorial(n - 2))
+       }
+ }
>
> # calling function cube with an argument
> factorial(5)
```

BUILT-IN FUNCTIONS

In r language, the built-in functions are categorized as the following:

- Math function

- Character function

- Statistical probability

- Other statistical functions

Everything in R is done through functions.[8] In the following, we are referring to numeric and character functions that are commonly used in creating variables. The list is given below:

Numeric Functions

Function	Its description
abs(x)	It returns absolute value of input x
sqrt(x)	It returns square root of input x
ceiling(x)	It returns the small integer which is larger than
floor(x)	or equal to x
trunc(x)	It returns the large integer, which is smaller
round(x, digits=n)	than or equal to x
signif(x, digits=n)	It returns the truncate value of input x
It is signif(3.475, digits=2) is 3.5	It returns round value of input x
cos(x), sin(x), tan(), acos(x), cosh(x),	It returns cos(x), sin(x), tan(x) value of input x
cosh(x), etc.	It is natural logarithm. It returns the natural
log(x)	logarithm of input x
log10(x)	It returns common logarithm of input x
exp(x)	It returns exponent (e^x)

Example of each of the following:

```
> x <- print(abs(2))
[1] 2
> x <- print(sqrt(5))
[1] 2.236068
> x <- print(ceiling(4))
[1] 4
> x <- print(floor(4))
[1] 4
> x<- c(2.2,6.56,10.11)
> print(trunc(x))
[1]  2  6 10
> x=2.456
> print(round(x,digits=2))
[1] 2.46
> x=2.4568
> print(round(x,digits=3))
[1] 2.457
> x<- 3
> print(cos(x))
[1] -0.9899925
> print(sin(x))
[1] 0.14112
> print(tan(x))
[1] -0.1425465
> x <- 3
> print(log(x))
[1] 1.098612
> x <- 3
> print(log10(x))
[1] 0.4771213
> x <- 3
> print(exp(x))
[1] 20.08554
```

Character Functions

The functions which are useful with characters and their manipulation are known as character functions in R.[9] We will discuss the following character functions:

- is.character()

- as.character()

- substr()

- grep()

- sub()

- strsplit()

- paste()

- toupper()

- tolower()

Let us discuss the above functions in detail.

Function	Description
is.character()	It checks if the given values (or a vector of values) are character/string or not
as.character()	It checks the input value to be converted into a character
substr(x, start=n1, stop=n2)	It extracts a substring from a given string object (or a vector containing string objects) based on the starting and the ending point x <- "abcdef"
grep(pattern, x, fixed=FALSE, ignore.case=FALSE,) It search for pattern in x.	It searches for the same specific pattern in the specified string and returns output accordingly. If fixed = "FALSE" then the pattern is a regular expression. If the fixed = "TRUE," then the pattern is a text string. The output will be the index of character in each element of the given string that matches the particular pattern
sub(pattern, replacement, , fixed=FALSE, x, ignore.case =FALSE) .	It checks for a specific character/substring in the given vector and replaces the first occurrence of it according to the input provided. Find the pattern in x and replace it with the replacement text. If you make fixed="FALSE," then pattern is a regular expression If fixed = T, then the pattern is a text string
strsplit(x, split) strsplit("abc", "")	It split the elements of character vector x at split It returns three element vectors "a," "b," "c"
paste(..., sep="")	It concatenates strings after using sep string to separate them paste("x",1:3,sep="") It returns c("x1", "x2" "x3") paste("x",1:3,sep="M") It returns c("xM1","xM2" "xM3") paste("Today is", date())
toupper(x)	It convert string in uppercase
tolower(x)	It convert string in lowercase

Statistical Probability Functions

The below table describes functions related to probability distributions. For random number generators, you can use set.seed(1234) or some other integer to create pseudo-random numbers.

Function	Description

dnorm(x)	It is normal density function (by default m=0 sd=1)
	# plot the standard normal curve
	x <- pretty(c(−3,3), 30)
	y <- dnorm(x)
	plot(x, y, type='l', xlab="Simple Deviate," ylab="Density", yaxs="i")

pnorm(q)	It is a cumulative normal probability for q (area under the normal curve to the left of q) pnorm(1.96) is 0.975
qnorm(p)	It is a normal quantile The value at p percentile of normal distribution qnorm(.9) is 1.28 # 90th percentile
rnorm(n, m=0,sd=1)	It is n random normal deviates with mean m and standard deviation sd #50 random variates with the mean=50, sd=10 x <- rnorm(50, m=50, sd=10)

dbinom(x, size, probability)
pbinom(q, size, probability)
qbinom(p, size, probability)
rbinom(n, size, probability)
	# binomial distribution where the size is the sample size
	and probability of a heads (pi)
	# probability of 0 to 5 heads of fair coin out of 10 flips
	dbinom(0:5, 10, .5)
	# probability of 5 or less heads of fair coin out of 10 flips
	pbinom(5, 10, .5)

dpois(x, lambda)
ppois(q, lamda)
qpois(p, lamda)
rpois(n, lamda)
	Poisson distribution with m=std=lamda
	#probability of 0, 1, or 2 events with lamda=4
	dpois(0:2, 4)
	# probability of at least three events with lamda=4
	1- ppois(2,4)

dunif(x, minimum=0, maximum=1) punif(q, minimum=0, maximum=1) qunif(p, minimum=0, maximum=1) runif(n, minimum=0, maximum=1)	Uniform distribution follows the same pattern as the normal distribution above #10 uniform random variates x <- runif(10)

Other Statistical Functions

Other useful statistical functions are listed in the following table. Each has a na.rm option to remove missing values before calculations. Otherwise, the missing values will lead to a missing result. An object can be a numeric vector or a data frame.

mean(x, trim=0 na.rm=FALSE)	Mean of object x # trimmed mean, removing the missing values and # 5 percent of the highest and lowest scores MX <- mean(x,trim=.05,na.rm=TRUE)
sd(x)	Standard deviation of object(x). also look at var(x) for variance and mad(x) for the median absolute deviation
median(x)	Median
quantile(x, probs)	Quantiles where x is the numeric vector quantiles are desired and probs is a numeric vector with probabilities in [0,1] # 30th and 84th percentiles of x y <- quantile(x, c(.3,.84))
range(x)	Range
sum(x)	Sum
diff(x, lag=1)	Lagged differences, with lag indicating which lag to use
min(x)	Minimum
max(x)	Maximum
scale(x, center=TRUE, scale=TRUE)	Column center or standardize a matrix
seq(from, to, by)	It generates a sequence indices <- seq(1, 12,2) # it indices is c(1, 3, 5, 7, 9)
rep(x, ntimes)	repeat x n times y <- rep(1:3, 2) # y is c(1, 2, 3, 1, 2, 3)
cut(x, n)	It divide continuous variable in factor with n levels y <- cut(x, 5)

R STRINGS

Any value encoded in a pair of single quotes (' ') and double quotes (" ") in R programming is called a "string."[10] Inside R, it wraps all strings in double quotes, even if you use single quotes to create them. In this chapter, we will learn about the basic concepts that R provides on strings.

Text data are stored in character vectors or character fields. It is important to remember that each element of a character vector represents the entire string, not just a single character. In R programming, "string" is a random term that is used as an "element of a character vector."

As you know, character vectors are required to be created using the c function. You can use single and double quotes around any string as long as they match, although double quotes are considered more common. Following is a simple example of how to use them.

The following are the rules to use string in R:[11]

- The string that starts with a single quote needs to end with a single quote. While you can put double quotes through the Escape Sequence('\'), single quote can also become a part of the string. For example: 'your', 'your"s', 'your\'s' are valid string in R.

- The string that starts with a double quote needs to end with a double quote. However, you can put single quotes, and through the Escape Sequence('\'), double quote can also become a part of the string. For example, your', 'your"s', 'your\'s' are also valid string in R.

String Basics

You can create strings either using single quotes or double quotes. Unlike other languages, there is no other difference in behavior. We recommend always using ", unless you want to create a string that contains multiple ".

Example:

```
> string1 <- "This is a string"
> string2 <- 'If I want to include a "quote"
inside a string, I use single quotes
> print(string1)
[1] "This is a string"
> print(string2)
[1] "If I want to include a \"quote\" inside a
string, I use single quotes"
```

If you forget to close a quote """, ' ', you will see +, the continuation character is as follows:

```
> "This is a string without closing quote
+
+
+
+
+
+
```

If this happens to you, press Escape and try again.

As you know, character vectors are required to be created using the c function. You can use single and double quotes around any string as long as they match, although double quotes are considered more common. Following is a simple example of how to use them,

c() Function in R

It combines values into a vector or list using the c() function in R. You can find the basic of R programming syntax of the c() function below.

Syntax:

```
c(value1, Val
ue2)
```

```
> c (
+       "Lorem ipsum dolora sit amet, consecteture
adipiscing elit.
+. ",
+       'Proin vestibulum diam posuere consequat
eleifend'
+ )
[1] "Lorem ipsum dolora sit amet, consecteture
adipiscing elit.\n."
[2] "Proin vestibulum diam posuere consequat
eleifend"
```

This function has two applications in R programming language:

1. Combine numeric value in vector using c() function

2. Combine variable and numeric values

Let's discuss each of them:

1. With numeric function: We simply have to specify if all values want to add new vector within the c function. We have to separate these values by a comma.

Example:

```
# Applying c() function on list
> x1 <- c(1, 2,  3, 8, 8, 4, 6)
> print(x1)
[1] 1 2 3 8 8 4 6
```

2. With variable and numeric values

We can append new values to the vector as shown in the following R code:

Example:

```
# Concatenate variable & values
> x2 <- c(x1, 5, 7, 1)
> print(x2)
 [1] 1 5 3 8 8 4 6 5 7 1
```

BUILT-IN STRING FUNCTION IN R

Base R includes many functions for working with strings, but we avoid them because they can be inconsistent, making them difficult to remember. Instead, we'll use functions from stringr. These have more intuitive names and all start with str_. For example, str_length() tells you the number of characters in a string.

We need to install and load the stringr R package.

```
# Install stringr R package
> install.packages("stringr")
WARNING: Rtools is a required to build R packages but
it is not currently installed. Please download and
install appropriate version of Rtools before
proceeding,
http://cran.rstudio.com/bin/windows/Rtools/
Installing package into 'C:/Users/Dell/AppData/
Local/R/win-library/4.2'
(as 'lib' is unspecified)
trying URL 'http://cran.rstudio.com/bin/windows/
contrib/4.2/stringr_1.4.1.zip'
Content type 'application/zip' length 218860 bytes
(213 KB)
downloaded 213 KB
package 'stringr' successfully unpacked and MD5 sums
checked
```

```
The downloaded binary packages are in
  C:\Users\Dell\AppData\Local\Temp\RtmpGiiChs\
downloaded_packages
> library("stringr")
> str_length(c("a", "R for data science", NA))
[1]  1 18 NA
> str_length(c("a", "R is a programming language for
data science", NA))
[1]  1 44 NA
```

Package "stringr"

1. case

2. invert_match

3. modifiers

4. stringr-data

5. str_c

6. str_conv

7. str_count

8. str_detect

9. str_dup

10. str_extract

11. str_flatten

12. str_glue

13. str_length

14. str_locate

15. str_match

16. str order

17. str_pad

18. str_remove

19. str_replace

20. str_replace_na

21. str_split

22. str_starts

23. str_sub

24. str_subset

25. str_trim

26. str_trunc

27. str_view

28. str_wrap word

Let's discuss some of them:

- str_to_upper(string, locale = "en")

- str_to_lower(string, locale = "en")

- str_to_title(string, locale = "en")

- str_to_sentence(string, locale = "en")

Example:

```
> language <- "R is a programming language"
> str_to_upper(language)
[1] "R IS A PROGRAMMING LANGUAGE"
> str_to_lower(language)
[1] "r is a programming language"
> str_to_title(language)
[1] "R Is A Programming Language"
> str_to_sentence("R is a programming language")
[1] "R is a programming language"
> # Locale matters!
> str_to_upper("i") # In English
[1] "I"
> str_to_upper("i", "tr") # In Turkish
[1] "İ"
```

- str_c

Example:

```
> str_c("Letter; ", letters)
 [1] "Letter; a" "Letter; b" "Letter; c"
"Letter; d"
 [5] "Letter; e" "Letter; f" "Letter; g"
"Letter; h"
 [9] "Letter; i" "Letter; j" "Letter; k"
"Letter; l"
[13] "Letter; m" "Letter; n" "Letter; o"
"Letter; p"
[17] "Letter; q" "Letter; r" "Letter; s"
"Letter; t"
[21] "Letter; u" "Letter; v" "Letter; w"
"Letter; x"
[25] "Letter; y" "Letter; z"
> str_c("Letter", letters, sep = "- ")
 [1] "Letter - a" "Letter- b" "Letter-  c"
"Letter- d"
 [5] "Letter: e" "Letter- f" "Letter- g"
"Letter: h"
 [9] "Letter- i" "Letter- j" "Letter- k"
"Letter- l"
[13] "Letter-  m" "Letter- n" "Letter- o"
"Letter- p"
[17] "Letter-  q" "Letter- r" "Letter- s"
"Letter- t"
[21] "Letter- u" "Letter:-v" "Letter- w"
"Letter:-x"
[25] "Letter- y" "Letter- z"
> str_c(letters, " is for", "...")
 [1] "a is for..." "b is for..." "c is for..."
 [4] "d is for..." "e is for..." "f is for..."
 [7] "g is for..." "h is for..." "i is for..."
[10] "j is for..." "k is for..." "l is for..."
[13] "m is for..." "n is for..." "o is for..."
[16] "p is for..." "q is for..." "r is for..."
[19] "s is for..." "t is for..." "u is for..."
[22] "v is for..." "w is for..." "x is for..."
[25] "y is for..." "z is for..."
> str_c(letters[-26], " it comes before ",
letters[-1])
 [1] "a comes before b" "b comes before c"
```

```
 [3] "c comes before d" "d comes before e"
 [5] "e comes before f" "f comes before g"
 [7] "g comes before h" "h comes before i"
 [9] "i comes before j" "j comes before k"
[11] "k comes before l" "l comes before m"
[13] "m comes before n" "n comes before o"
[15] "o comes before p" "p comes before q"
[17] "q comes before r" "r comes before s"
[19] "s comes before t" "t comes before u"
[21] "u comes before v" "v comes before w"
[23] "w comes before x" "x comes before y"
[25] "y comes before z"
> str_c(letters, collapse = "")
[1] "abcdefghijklmnopqrstuvwxyz"
> str_c(letters, collapse = ", ")
[1] "a, b, c, d, e, f, g, h, i, j, k, ll, m, n,
o, p, q, r, s, t, u, v, w, x, y, z"
> # Missing inputs give missing outputs
> str_c(c("a", NA, "b"), "-d")
[1] "a-d" NA        "b-d"
> # Use str_replace_NA to the display literal
NAs:
> str_c(str_replace_na(c("aa", NA, "b")), "-d")
[1] "aa-d"  "NA-d" "b-d"
```

- str_count

Example:

```
> fruit <- c("apple", "banana", "pear",
"pineapple")
> str_count(fruit, "a")
[1] 1 3 1 1
> str_count(fruit, "p")
[1] 2 0 1 3
> str_count(fruit, "e")
[1] 1 0 1 2
> str_count(fruit, c("a", "i", "p", "p"))
[1] 1 1 1 3
> str_count(c("a.", "...", ".a.a"), ".")
[1] 2 3 4
> str_count(c("a.", "...", ".a.a"), fixed("."))
```

- str_length

Example:

```
> library("stringr")
> str_length(letters)
 [1] 1 1 1 1 1 1 1 1 1 1 1 1 1 1 1 1 1 1 1 1 1 1 1
1 1 1
[26] 1
> str_length(NA)
[1] NA
> str_length(factor("ABC"))
[1] 3
> str_length(c("i", "like", "programming", NA))
[1]  1  4 11 NA
> # Two ways of representing a u with an umlaut
> u1 <- "\u00fc"
> u2 <- stringi::stri_trans_nfd(u1)
> # The print the same:
> u1
[1] "ü"
> u2
[1] "ü"
> # But have a different length
> str_length(u1)
[1] 1
> str_length(u2)
[1] 2
> # Even though they have the same number of
characters
> str_count(u1)
[1] 1
> str_count(u2)
[1] 1
```

String Manipulation
Concatenating strings: paste() function
You can combine many strings in R using the paste() function. It can take any number of arguments.

Example:

```
> a <- "You are "
> b <- 'learning '
```

```
> c <- "R Programing Language "
> print(paste(a,b,c))
[1] "You are  learning  R Programing Language "
> print(paste(a,b,c, sep = "-"))
[1] "You are -learning -R Programing Language "
> print(paste(a,b,c, sep = "", collapse = ""))
[1] "You are learning R Programing Language "
```

Formatting numbers and strings format() function

You can format numbers and strings using the format() function.

Example:

```
> # The total number of digits displayed. Last
digit rounded off.
> result <- format(13.6789, digits = 4)
> print(result)
[1] "13.68"
>
> # Display numbers in scientific notation.
> result <- format(c(3, 13.521), scientific =
TRUE)
> print(result)
[1] "3.0000e+00" "1.3521e+01"
>
> # The min number of digits to the right of the
decimal point.
> result <- format(12.42, small = 2)
> print(result)
[1] "12.42"
>
> # It format treats everything as a string.
> result <- format(5)
> print(result)
[1] "5"
>
> # The numbers are padded with blank in the
beginning for width.
> result <- format(15.7, width = 3)
> print(result)
[1] "15.7"
>
```

```
> # The left justify strings.
> result <- format("Hello world", width = 8,
justify = "l")
> print(result)
[1] "Hello world"
>
> # The justify string with center.
> result <- format("Hello World", width = 8,
justify = "c")
> print(result)
[1] "Hello World"
```

CHAPTER SUMMARY

In this chapter, we learned definition, usage, types, and user-defined functions in R along with examples. We also covered string in R in the same chapter; some built-in string functions to make coding easier.

NOTES

1. Functions in R – https://www.w3schools.in/r-programming/r-functions, accessed on October 7, 2022.
2. Function – https://www.w3adda.com/r-tutorial/r-function#r-function, accessed on October 7, 2022.
3. Function Component – https://www.tutorialspoint.com/r/r_functions.htm, accessed on October 7, 2022.
4. Argument in Functions – https://www.w3schools.com/r/r_functions.asp, accessed on October 7, 2022.
5. Function Types – https://www.geeksforgeeks.org/types-of-functions-in-r-programming/, accessed on October 7, 2022.
6. Infix Functions – https://colinfay.me/playing-r-infix-functions/, accessed on October 7, 2022.
7. Functions – https://www.w3schools.com/r/r_variables_global.asp, accessed on October 7, 2022.
8. Built-in Function – https://www.statmethods.net/management/functions.html, accessed on October 8, 2022.
9. Built-in Function – https://www.analyticssteps.com/blogs/character-functions-r, accessed on October 8, 2022.
10. String – https://www.w3schools.in/r-programming/strings/, accessed on October 8, 2022.
11. Rules of String – https://www.datacamp.com/tutorial/strings-in-r, accessed on October 8, 2022.

## CHAPTER 6

Lists and Arrays

IN THIS CHAPTER

➢ Introduction

➢ List Operation

➢ Double versus Single Parentheses

➢ and More

The concept of function and string was also discussed in the last chapter with their types and built-in functions. This chapter is all about the list and array in detail.

INTRODUCTION

Lists are R objects that contain elements of various types such as numbers, strings, vectors, and other lists. A list can also contain a matrix or a function as its elements. The list is created by using the list() function.

LIST OPERATIONS

A list is a generic object consisting of an ordered collection of objects.[1] These are one-dimensional, heterogeneous data structures. A list can be also a list of vectors, a list of matrices with characters, a list of functions, and so on.

A list is a vector with heterogeneous data elements. A list in R is cre ated using the list() function. R allows you to access the elements of a list

DOI: 10.1201/9781003358480-6

using an index value. In R, list indexing starts at 1 instead of 0 as in other programming languages.

Creating a List

To create a list in R, you need to use a function called "list()."[2] In other words, a list is a general vector containing other objects. To illustrate what the list looks like, here's an example. We want to build a list of employees with details. For this, we need attributes like id, employee name, and number of employees. We will create an R list using an example. You can create a list containing strings, numbers, vectors, and Booleans.

Example:

```
> #Here we assign different
> list_data <- list( " Red ", " R Programming ",
c(11,22,33), FALSE,  22.40, 51)
> print(list_data)
[[1]]
[1] "  Red  "
[[2]]
[1] "  R Programming  "
[[3]]
[1] 11 22 33
[[4]]
[1] FALSE
[[5]]
[1] 22.4
[[6]]
[1] 51
```

A list is a general vector containing other objects. For example, the following variable x is a list containing copies of the three vectors n, s, and b and the numeric value 3.

Example:

```
> n = c(2, 3, 5)
> s = c("aa", "bb", "cc", "dd", "ee")
> b = c(TRUE, FALSE, TRUE, FALSE, FALSE)
> x = list(n, s, b, 3)   # x contains copies of n,
s, b
```

```
> print(n)
[1] 2 3 5
> print(s)
[1] "aa" "bb" "cc" "dd" "ee"
> print(b)
[1]   TRUE FALSE  TRUE FALSE FALSE
> print(x)
[[1]]
[1] 2 3 5

[[2]]
[1] "aa" "bb" "cc" "dd" "ee"

[[3]]
[1]   TRUE FALSE  TRUE FALSE FALSE

[[4]]
[1] 3
```

Another example:

```
> # R program to create a simple list
> # The attributes is a numeric vector
> # containing all the employee IDs which is
created
> # using the command here
> emp_Id = c(1, 2, 3, 4)
> # The next second attribute is the employee name
> # which is created by using this line of code
here
> # which is the character vector
> emp_Name = c("Bedi", "Deep", "Lav", "Jas")
> # Then third attribute is the number of
employees
> # which is a single numeric variable.
> numberOfEmp = 4
> # We combine all these three different
> # data types into a list
> # containing details of the employees
> # which can be done by using a list command
> empList = list(emp_Id, emp_Name, numberOfEmp)
> print(empList)
[[1]]
```

```
[1] 1 2 3 4
[[2]]
[1] "Bedi" "Deep" "Lav"   "Jas"
[[3]]
[1] 4
```

Naming List Elements in R

In this section, we will also learn to name R list elements with the help of an example. Let us create a list containing a vector, matrix, and list.

Example:

```
> data_list <- list(c(" Monday "," Tuesday "," 
Wednesday "), matrix(c(1, 2, 3, 4, 5, 6), nrow = 
3),list(" Yellow ", 12.3))
> names(data_list)
NULL
> data_list <- list(c(" Monday "," Tuesday "," 
Wednesday "), matrix(c(1, 2, 3, 4, 5, 6), nrow = 
3),list(" Yellow ", 12.3))
> names(data_list) <- c( "CCC", "DDD", "EEE")
> print(data_list)
$CCC
[1] " Monday "       " Tuesday "
[3] " Wednesday "

$DDD
      [,1] [,2]
[1,]    1    4
[2,]    2    5
[3,]    3    6

$EEE
$EEE[[1]]
[1] " Yellow "

$EEE[[2]]
[1] 12.3
```

Access to R List Elements

Let us now understand how to access the elements of lists in R programming. R creates a list containing a vector, a list, and a matrix. We will use the list we created in the previous section "data_list."

Example:

```
> v <- vector("character", 5)
> v[1] <- 'a'
> v[2] <- 'a'
> v[4] <- 'a'
> v
[1] "a" "a" ""  "a" ""
> # [1] "a" "a" ""  "a" ""
> c <- list(v, v)
> names(c) <- c("var1", "var2")
> c
$var1
[1] "a" "a" ""  "a" ""

$var2
[1] "a" "a" ""  "a" ""

> # $var1
> # [1] "a" "a" ""  "a" ""
>
> # $var2
> # [1] "a" "a" ""  "a" ""
>
> c["var1"][1]
$var1
[1] "a" "a" ""  "a" ""

> # $var1
> # [1] "a" "a" ""  "a" ""
>
> c$"var1"[1]
[1] "a"
> # [1] "a"
```

Another example: In order to access list elements, use index number, and in the case of a named list, elements can access using its name also.

```
> # Creating a List
> list <- list(c( "R", "Programming"),
+                 matrix(c(1:9), nrow = 3),
+                 list("Geek", 12.3))

> # Naming each element of the list
```

```
> names(list) <- c("This is a vector",
+                   "This is a_Matrix",
+                   "This is a list within the
list")

> # To access an element of the list.
> print(list[1])
$`This is a vector
[1] "R"            "Programming"


> #  To access the third element.
> print(list[3])
$'This is a list within the list
$'This is a list within the list[[1]]
[1] "Geek"

$`This is a list within the list[[2]]
[1] 12.3
```

A list is actually a vector but it is different in comparison to the other types of vectors which we are using in this class.

- Other vectors are atomic vectors.

- A list is a type of vector called a recursive vector.

There are some more operations you can add on the list.

INDEXING LISTS

For vectors, arrays, and matrices, we saw that their indexing was very similar except for the dimensions. However, the list is very different. Note that unlike a typical vector, this one prints in multiple parts. This also allows to help with indexing, as we'll see below. There is another easy way to create the same list.

Example:

```
>  list = c( "Monday", "Tuesday",  "Wednesday",
"Thursday",  "Friday", "Saturday", "Sunday" )
#index using logical vector
```

```
> print(list[2])
[1] "Tuesday"
> print(list[5])
[1] "Friday"
> print(list[7])
[1] "Sunday"
```

SLICING THE LIST

We load a slice of the list with one square bracket "[]" operator. Next is a cut containing the second term x, which is a copy of s.

Example:

```
> x[2]
[[1]]
[1] "aa" "bb" "cc" "dd" "ee"
```

With an index vector, we can retrieve a slice with multiple members. Following is a slice containing the second and fourth members of x.

Example:

```
> x[c(2, 4)]
[[1]]
[1] "aa" "bb" "cc" "dd" "ee"

[[2]]
[1] 3 [[2]]
```

Another example:

```
> # here we create a list
> list_data <- list(c(" R ", " is ", " Programming
Language"),
+                   matrix(c(1:9), nrow = 3),
+                   list("Geek", 12.3))
> # Naming each element of the list
> names(data_list) <- c("This_is_vector",
+                   "this_is_matrix",
+                   "This_is_a_list_in_a_list")
```

```
> # To access the first element of the list.
> print(data_list[1])
$This_is_vector
[1] " Monday "      " Tuesday "
[3] " Wednesday "
> # To access the third element.
> print(data_list[3])
$This_is_a_list_in_a_list
$This_is_a_list_in_a_list[[1]]
[1] " Yellow "
$This_is_a_list_in_a_list[[2]]
[1] 12.3
> # Access a list element by element name.
> print(data_list$This_is_a_Matrix)
NULL
```

There are some predefined R lists. Following are lists for letters and month names are predefined such as

- letters
- LETTERS
- month.abb
- month.name

Run the above predefined list in R console.

Letters

Here is an example of the Letters:

```
> letters
 [1] "a" "b" "c" "d" "e" "f" "g" "h" "i"
[10] "j" "k" "l" "m" "n" "o" "p" "q" "r"
[19] "s" "t" "u" "v" "we" "x" "y" "z"
```

LETTERS

Here is an example of the LETTERS:

```
> LETTERS
 [1] "A" "B" "C" "D" "E" "F" "G" "H" "I"
```

```
[10] "J" "K" "L" "M" "N" "O" "P" "Q" "R"
[19] "S" "T" "U" "V" "W" "X" "Y" "Z"
```

month.abb

Here is an example of the month.abb:

```
> month.abb
 [1] "Jan" "Feb" "Mar" "Apr" "May" "Jun"
 [7] "Jul" "Aug" "Sep" "Oct" "Nov" "Dec"
```

month.name

Here is an example of the month.name:

```
> month.name
 [1] "January"   "February"   "March"
 [4] "April"     "May"        "June"
 [7] "July"      "August"     "September"
[10] "October"   "November"   "December"
```

ADDING, DELETING, AND UPDATING ELEMENTS OF A LIST

In R language, a new element can be added to the list; the existing element can be deleted or updated.

Example:

```
> # Creating a List
> list <- list(c(" R ", " Programming "),
+                matrix(c(1:6), nrow = 2),
+                list("Programming", 12.2))

> # Naming each element of the list
> names(list) <- c("This is a vector",
+                   "This is a Matrix",
+                   "This is a list within the
list")

> # To add a new element.
> list[4] <- "New element"
> print(list)
$'This is a vector
[1] " R "              " Programming "
```

```
$'This is a Matrix'
      [,1]  [,2]  [,3]
[1,]    1     3     5
[2,]    2     4     6


$'This is a list within the list
$'This is a list within the list[[1]]
[1] "Programming"


$'This is a list within the list[[2]]
[1] 12.2



[[4]]
[1] "New element"


> # To remove the last element.
> list[4] <- NULL
>
> # To print the 4th Element.
> print(list[4])
$<NA>
NULL


> # To update the 3rd Element.
> list[3] <- "updated element"
> print(list[3])
$'This is a list within the list
[1] "updated element"
```

CONVERTING A LIST TO VECTOR

In order to perform arithmetic operations, lists can be converted to vectors
using the unlist() function.

Example:

```
> # Firstly, create lists.
> list1 <- list(1:5)
> print(list1)
[[1]]
[1] 1 2 3 4 5


> list2 <-list(11:15)
```

```
> print(list2)
[[1]]
[1] 11 12 13 14 15
> # Now, convert the lists to vectors.
> v1 <- unlist(list1)
> v2 <- unlist(list2)
> print(v1)
[1] 1 2 3 4 5
> print(v2)
[1] 11 12 13 14 15
> # Now add the vectors
> result_vector <- v1+v2
> print(result_vector)
[1] 12 14 16 18 20
```

DOUBLE VERSUS SINGLE PARENTHESES

Single and double square brackets are used as indexing operators in the R programming language.[3] Both of these operators are used to refer to components of R objects either as a subset belonging to the same data type or as an element.

[ ]	[[ ]]
It used for singular indexing	It used for recursive indexing
It access list within a list	It access elements within a list
It may return a single element from the list	It also returns a single element from the list
It allows indexing by vectors	It allows indexing by integers or characters

LIST LENGTH

To find how many items a list has, use the length() function.

Example:

```
> this_list <- list("PHP", "HTML", "Ruby")
> length(this_list)
[1] 3
```

Check If Item Exists

If you want to find out if a specified item is present in a list, use the %in% operator.

Example:

```
To check if "Ruby" is present in the list,
> this_list <- list("PHP", "Ruby", "R")
> "Ruby" %in% this_list
[1] TRUE
```

ADD LIST IN ITEMS

To add the item to the end of list, use the append() function.

Example:

```
To add "R" to the list,
this_list <- list("R", "PHP", "Ruby")
append(this_list, "PHP")
```

Also to add an item to the right of a particular index, add "after=index number" in the append() function.

Example: To add "Ruby" to the list after "Ruby" (index 2).

```
> this_list <- list("PHP", "Ruby", "R")
> append(this_list, "Ruby", after = 2)
[[1]]
[1] "PHP"
[[2]]
[1] "Ruby"
[[3]]
[1] "Ruby"
[[4]]
[1] "R"
```

REMOVE LIST ITEMS

You can also remove list items. The below example creates a new, updated list without a "Ruby" item:

Example: To remove "apple" from the list.

```
>   this_list <- list("PHP", "Ruby", "R")
> new_list <- this_list[-1]
> # Print the new_ list
> new_list
```

```
[[1]]
[1] "Ruby"

[[2]]
[1] "R"
```

RANGE OF INDEXES IN LIST

You can also specify a range of indexes just by specifying where to start and where to end the range by using the : operator.

Example: It returns the second, third, fourth, and fifth item.

```
> this_list <- list("PHP", "Ruby", "R")
> (this_list)[2:5]
[[1]]
[1] "Ruby"

[[2]]
[1] "R"

[[3]]
NULL

[[4]]
NULL
```

LOOP THROUGH A LIST

You can loop through the list of items by using a for loop.

Example: It prints all items in the list one by one.

```
> this_list <- list("PHP", "Ruby", "R")
> (this_list)[2:5]
[[1]]
[1] "Ruby"

[[2]]
[1] "R"

[[3]]
NULL

[[4]]
NULL
```

Join Two Lists

There are various methods to join, or concatenate, more than two lists in R. The common way is to use c() function, which combines two elements together:

Example:

```
> list1 <- list("a", "b", "c")
> list2 <- list(1,2,3)
> list3 <- c(list1,list2)
> print(list1)
[[1]]
[1] "a"

[[2]]
[1] "b"

[[3]]
[1] "c"

> print(list2)
[[1]]
[1] 1

[[2]]
[1] 2

[[3]]
[1] 3

> print(list3)
[[1]]
[1] "a"

[[2]]
[1] "b"

[[3]]
[1] "c"

[[4]]
[1] 1
```

```
[[5]]
[1]  2

[[6]]
[1]  3
```

CHAPTER SUMMARY

You have learned about R array along with its creation, updating, adding indexing, and manipulating array in an array with examples, with the creation and addition of data in the list.

NOTES

1. List Operation – https://www.geeksforgeeks.org/r-lists/, accessed on October 7, 2022.
2. Creating R – https://www.tutorialspoint.com/r/r_lists.htm, accessed on October 7, 2022.
3. List in R – https://www.w3schools.com/r/r_lists.asp/, accessed on October 8, 2022.

## CHAPTER 7

Data Structures

IN THIS CHAPTER

- ➤ Introduction
- ➤ Types of Data Structure
- ➤ Vector
- ➤ Matrix
- ➤ Data Fame
- ➤ and more

The last chapter was about lists and arrays that we create, update, delete list, and array with various built-in functions. In this chapter, we will study different types of data structures in R programming. We will also understand their use and implementation with the help of examples.

INTRODUCTION

In any programming language, if you have done programming, you have to use different variables to store different data.[1] In addition, variables are reserved in a memory location for storing values. This also means that once you create a variable, you reserve a certain area in memory. Furthermore, data structures are the only way to organize data so that they can be used efficiently on a computer.

DOI: 10.1201/9781003358480-7

As far as we can see, unlike other programming languages like C and Java, R does not have variables declared as any data type. Furthermore, variables are named with R objects; the knowledge form of the R object becomes the data type of the variable.

There are many types of R objects. The most frequently used are the following:

- Vector

- Matrix

- Array

- Lists

- Data frames

Now that we know how to talk to R through the script editor or the console, we want to use it for something more than just adding numbers. For this, we need to know more about R syntax. Below is a sample script highlighting many different "parts of speech" for R (syntax):

- comments # and with how they are used to document a function and its contents

- variables and functions

- assignment operator <-

- = for arguments in functions

We know that variables are like segments, and so far, we've seen that segments are filled with a single value. Even when the number was created, the result of the mathematical operation was a single value. Variables can store more than just one value; they can store a large number of different data structures. These include not limited to vectors (c), factors, matrices (matrix), data frames (data.frame), and lists (list).

VECTOR

A vector is a data structure in R. It contains an element of the same type. Data types can be Boolean, integer, double, character, complex, or raw. Type of a vector can be checked using the typeof() function. Another property of a vector is its length. The various numbers of elements in the vector that can be checked using the length() function. Vectors are R's most basic data objects, and there are six types of atomic vectors. They are Boolean, integer, double, complex, character, and raw.

Single Element Vector

Even if you write only one value to R, it becomes a vector of 1 length and belongs to one of the vector types.

How to Create Vector in R?

Vectors are created using the c() function. Since a vector must have elements of the same type, the function will try and coerce elements to the similar type if they are different. Coercion is from low to high types and from logical to integer to double to character.

```
> x <- c(11, 55, 44, 99, 00)
> typeof(x)
[1] "double"
> length(x)
[1] 5
>
> x <- c(11, 54, TRUE, "R Programming")
> x
[1] "11"              "54"
[3] "TRUE"            "R Programming"
> typeof(x)
[1] "character"
```

A vector is the common and basic data structure in R and is pretty much the workhorse of R. It is basically only a collection of values, mainly either numbers.

Atomic Vectors

There are four types of R atomic vectors:

- Character data type

- Logical data type

- Numeric data type

- Integer data type

There are the following ways to access the element of a vector. Vectors are created using the c() function. Since the vectors must have elements of the same type, this function will try to force elements of the same type if they are different. Coercion is from lower to higher types and from Boolean to integer to double to character.

```
> x <- c(11, 55, 44, 99, 00)
> typeof(x)
[1] "double"
> length(x)
[1] 5
>
> x <- c(11, 54, TRUE, "R Programming")
> x
[1] "11"              "54"
[3] "TRUE"            "R Programming"
> typeof(x)
[1] "character"
```

Types of Vectors

Vectors are of different kinds which are used in R. The following are some of the types of vectors:

Numeric Vectors

These vectors contain numeric values such as integer, float, and so on.

Example:

```
> # R program to create numeric Vectors
> # creation of vectors using c() function.
> v1 <- c(14, 15, 16, 17)
> # display type of vector
> typeof(v1)
[1] "double"
> # by using 'L' you can specify that you want
integer values.
> v2 <- c(11L, 44L, 22L, 55L)
> # display type of vector
> typeof(v2)
[1] "integer"
```

Character Vectors
Character vectors contain alphanumeric values and special characters.

Example:

```
> # R program to create Character Vectors
> # by default numeric values
> # are converted into characters
> v1 <- c(' R Programming ', '2', 'Vectors', 57)
>
> # Displaying type of vector
> typeof(v1)
[1] "character"
```

Logical Vectors
Logical vectors contain Boolean values, such as TRUE, FALSE, and NA for null values.

Example:

```
> # R program to create Logical Vectors
> # Creating logical vector
> # using c() function
> v1 <- c(TRUE, FALSE, TRUE, NA)
>
> # Displaying type of vector
> typeof(v1)
[1] "logical"
```

CREATION OF VECTOR IN R

If you want to create a vector in R language, there are three different ways.[2]

Create Vector Using c() Function

The c() function combines values together and we can use the c() to combine the values and create a vector. In the following code, we will use c() to create a vector of double values. Three different ways are there to create it, that is, c() function, seq() function, and : operator.

Example:

```
> x <- c(11, 44,99, 116,125)
> cat(x)
11 44 99 116 125
```

Create Vector Using seq() Function

seq() function creates a sequence of values and can use the seq() function to create a vector with the sequence of the values. In the following program, we will use the seq() function to create a vector of two values.

Example:

```
> x <- seq(0, 20, length.out=4)
> cat(x)
0 6.666667 13.33333 20
```

Create Vector Using: Operator

By using this operator, we can create a vector with continuous values. In the following program, we will use : operator to create a vector.

Example:

```
> x <- 1:12
> cat(x)
> #result
1 2 3 4 5 6 7 8 9 10 11 12
```

R MATRIX

Matrix is a two-dimensional data structure in R programming.[3] Matrix is the same as vector but it contains the dimension attribute. All attributes of an object can check with the attributes() function.

Creating a Matrix in R

Matrix can create using the matrix() function. The dimension of the matrix can be defined by passing values for arguments nrow and ncol.

Example:

```
> matrix(1:9, nrow = 3, ncol = 3)
     [,1] [,2] [,3]
[1,]    1    4    7
[2,]    2    5    8
[3,]    3    6    9
```

You can name the rows and columns of the matrix during creation by passing a two-element list to the argument dimnames.

Example:

```
> x <- matrix(1:9, nrow = 3, dimnames =
list(c("X","Y","Z"), c("A","B","C")))
> colnames(x)
[1] "A" "B" "C"
> rownames(x)
[1] "X" "Y" "Z"
```

It can be assessed by using the colname() and rownames() function. There are another ways of creating matrix by using cbind() and rbind() functions.

Example:

```
> cbind(c(11, 22, 33),c(44, 55, 66))
     [,1] [,2]
[1,]   11   44
[2,]   22   55
[3,]   33   66
>
> rbind(c(01, 02, 03),c(04, 05, 06))
     [,1] [,2] [,3]
[1,]    1    2    3
[2,]    4    5    6
```

Access Elements of a Matrix

We can access matrix elements using the square bracket [indexing method. The elements can access as var[row, column]. In the following, rows and columns are vectors.

Example:

```
> x <- matrix(1:9, nrow = 3, dimnames =
list(c("X","Y","Z"), c("A","B","C")))
> > x[c(1,2),c(2,3)]     # select rows 1 & 2 and
columns 2 & 3
Error: unexpected '>' in ">"
>   x[c(1,2),c(2,3)]     # select rows 1 & 2 and
columns 2 & 3
  BC
X 4 7
Y 5 8
> [,1] [,2]
```

While indexing, you can also find the element of matrix.

Example:

```
> x
  A B C
X 1 4 7
Y 2 5 8
Z 3 6 9
> x[1:4]
[1] 1 2 3 4
```

DATA FRAME

A data frame is a two-dimensional array structure in which each column contains the values of one variable, and a row contains a set of values from each column.[4] The following are the characteristics of the data frame.

- Column names should not be empty.

- The row names should be unique.

- The data stored in a data frame should be numeric, factor, or character.

- Each column contains the same number of data items.

CHAPTER SUMMARY

In this chapter, we discussed each of these components to better understand data structures in R. A data structure is defined as a specific form of data organization and storage. R programming supports five basic types of data structures, namely, vector, matrix, list, data frame, and factor.

NOTES

1. DS – https://www.tutorialkart.com/r-tutorial/r-create-vector/, accessed on October 10, 2022.
2. Vector – https://www.geeksforgeeks.org/r-vector/, accessed on October 10, 2022.
3. Matrix – https://www.tutorialspoint.com/r/r_matrices.htm, accessed on October 10, 2022.
4. Data Frame – https://www.tutorialspoint.com/r/r_data_frames.htm, accessed on October 10, 2022.

## CHAPTER 8

Error Handling and File Handling

IN THIS CHAPTER

- ➤ Introduction
- ➤ Handling Error Functions
- ➤ Manipulation of Conditions in R
- ➤ Debugging in R Programming
- ➤ File Handling in R
- ➤ and More

In this chapter, we will introduce some basic error-handling functions in R. If you are working with other programming languages, you may experience in how to use try, catch, and finally block to deal with possible errors during development. Similarly, R provides error-handling operations in its functions.

INTRODUCTION

Errors often occur when code is used that it is not intended to be used. For example, adding two strings together produces the following error.

```
> " You are learning " + "R Programming"
```

DOI: 10.1201/9781003358480-8

```
Error in " You are learning " + "R Programming": non-
numeric argument to binary operator
```

The + operator is a function that takes two numbers as arguments and finds their sum. In the example, we pass the string named > "You are learning" + "R Programming" where neither "You are learning" nor "R Programming" is the number, the R interpreter will always produce an error. Errors will stop the execution of your program, and they will print an error message to the R console "Error in 'You are learning' + 'R Programming': non-numeric argument to binary operator."

In R, there are two constructs which are related to errors, such as warnings and messages.

Warning

Warnings are meant to define that something seems to have gone wrong in your code that should be inspected. Here is an example of a warning being generated.

```
> as.numeric(c(" 1 ", " 2 ",  " three "))
[1]   1   2 NA
Warning message:
NAs introduced by coercion
```

The as.numeric() attempts to convert each string in c("1," "2," "three") into a number; however, it is impossible to convert "three," so a warning is generated. The execution of the code is not halted, NA is produced for "seven" instead of a number.

Message

As messages simply print to the R console, using this, they are generated by a mechanism that is similar to how errors and warnings are generated. In the following, a small function will generate a message.

Example:

```
> function_var <- function(){
+       message("This is a message.")
+ }
>
> function_var()
This is a message.
```

HANDLING ERROR FUNCTIONS

Error handling is the process by which we deal with unwanted or anomalous errors that may cause a program to terminate abnormally during its execution. There are two commonly used ways in R that we can implement an error-handling mechanism either we can directly call functions like stop() or warning() or we can use error options like "warn" or "warning. expression." Basic functions can be used to handle errors in your code.

- stop(...): It halts the evaluation of the current statement and generates a message argument. It controls return to the top level.

- waiting(...): Its evaluation depends on the value of the error option warn. If the value is (–) negative, then it is ignored, if the value is 0 (zero), they are stored and printed after the top-level function completes its execution.

- tryCatch(...): It helps to calculate the code and assign the exceptions.

Here is the list of common errors that occurs in the R programming.

- Error message: Attempt to apply non-function

- Error message: Can't rename columns that don't exist

- Error message: Cannot allocate vector of size X Gb

- Error message: Continuous value supplied to discrete scale

- Error message: Discrete value supplied to a continuous scale

- Error message: "R" is an unrecognized escape in character string starting ""C:R"

- Error: gcom_point requircs following missing acsthctics: x, y

- Error message: Insufficient values in manual scale. X was needed but only Y was provided.

- Error message: Invalid input: date_trans works with objects of class Date only

- Error message: JAVA_HOME cannot be determined from the Registry

- Error message: Mapping should be created with 'aes()' or 'aes_()'.

- Error message: object X not found

- Error: stat_count() not be used with a y aesthetic.

- Error message: unexpected ';' in ";"

- Error message: unexpected '}' in X

- Error message: unexpected '=' in "="

- Error message: unexpected ')' in ")"

- Error message: unexpected 'else' in "else"

- Error message: unexpected input in X

- Error message: unexpected numeric constant in X

- Error message: unexpected SPECIAL in X

- Error message: unexpected string constant in X

- Error message: unexpected symbol in X

- Error message: X must only be used inside dplyr verbs

- Error in .Call.graphics: the invalid graphics state

- Error in .subset(x, j): the invalid subscript type 'list'

- Error in apply(data): dim(X) have a positive length

- Error in as.Date.numeric(X): "origin" must be supplied

- Error in as.POSIXlt.character(x, tz, …): the character string is not in a standard unambiguous format

- Error in barplot.default(X): the 'height' must be a vector or a matrix

- Error in charToDate(x): the character string is not in a standard unambiguous format

- Error in colMeans(x, na.rm = TRUE): the 'x' must be numeric

- Error in contrasts: the contrasts can be applied only to factors with two or more levels

- Error in cut.default: 'breaks' are not the unique

- Error in file(file, "rt"): they cannot open the connection

- Error in file(file, "rt"): the invalid 'description' argument

- Error in fix.by(by.x, x): 'by' specify a uniquely valid column

- Error in FUN(X[[i]], …): the object 'X' not found

- Error in FUN: the invalid 'type' (character) of argument

- Error in FUN: the invalid 'type' (list) of argument

- Error in hist.default(X): the 'x' must be numeric

- Error in if (NA) {: missing value where the TRUE/FALSE needed

- Error in library("Y"): there is no package called 'Y'

- Error in lm.fit(x, y, offset = offset, singular.ok = singular.ok, …): the 0 (non-NA) cases

- Error in lm.fit(x, y, offset = offset, singular.ok = singular.ok, …): the NA/NaN/Inf in 'x'

- Error in load("X.rds"): bad restore the file magic number (file may be corrupted) – no data loaded

- Error in max.item – min.item: the nonnumeric argument to binary operator

- Error in model.frame.default(Terms, new data, na.action = a.action, xlev = object$xlevels): "data" must be a data.frame, environment, or list

- Error in model.frame.default(Terms, new data, na.action = a.action, xlev = object$xlevels): factor X has new levels Y

- Error in names(X): 'names' attribute must be of the same length as the vector

- Error in Ops.data.frame(): only defined for the equally-sized data frames

- Error in parse(text): <text>:1:3: unexpected symbol

- Error in plot.new(): figure margins are too large

- Error in plot.window(…): need the finite 'xlim' the values

- Error in plot.xy(xy.coords(x, y), type = type, …): the plot.new has not been called yet

- Error in rbind(deparse.level, …): the numbers of columns of arguments do not match

- Error in read.table: more columns than the column names

- Error in read.table(file = file, header = header, sep = sep, quote = quote,: duplicate "row.names" are not allowed

- Error in rep(X): invalid 'times' of argument

- Error in rowSums & colSums: "x" must be an array of at least two dimensions

- Error in scan: line 1 did not have X elements

- Error in setwd(X): cannot change working directory

- Error in solve.default(X): Lapack routine dgesv: system is exactly singular

- Error in sort.int(x, na.last = na.last, decreasing = decreasing, …): "x" must be atomic

- Error in stripchart.default(x1, …): invalid plotting method

- Error in strsplit(X): non-character argument

- Error in unique.default(x, nmax = nmax): unique() applies only to vectors

- Error in UseMethod("predict"): no applicable method for "predict" applied to an object of class "c('double', 'numeric')"

- Error in X variable: $ operator is invalid for atomic vectors

- Error in X variable: argument is of length zero

- Error in X variable: arguments imply differing number of rows

- Error in X variable: attempt to select less than one element in get1index real

- Error in X variable: could not find function X

- Error in X variable: missing values are not allowed in subscripted assignments of data frames

- Error in X variable: incorrect number of dimensions

- Error in X variable: incorrect number of subscripts on matrix

- Error in X variable: invalid (do_set) left-hand side to assignment

- Error in X variable: invalid (NULL) left side of assignment

- Error in X variable: non-numeric argument to binary operator

- Error in X variable: object not interpretable as a factor

- Error in X: the object of type 'closure' is not subsettable

- Error in X variable: replacement has Y rows, data has Z

- Error in X variable: requires numeric/complex matrix/vector arguments

- Error in X variable: subscript out of bounds

- Error in X variable: target of assignment expands to non-language object

- Error in X variable: undefined columns selected

- Error in X variable: unused argument

- Error in X %*% Y variable: non-conformable arguments

- Error in xy.coords(x, y, xlabel, ylabel, log): "x" and "y" lengths differ

- geom_path: Each group consists of only one observation. Do you need to adjust the group aesthetic?

- Please select a CRAN mirror for use in this session

- Scale for "fill" is already present. Adding another scale for "fill," which will replace the existing scale.

- The following objects are masked from the "package:X"

- The following objects are masked from X

- Use of 'data$X' is discouraged. Use 'X' instead

- Warning message: cannot remove the prior installation of package 'X'

- Warning error message: "new data" had X rows but variables found have Y rows

- Warning error message: In cor(X): the standard deviation is zero

- Warning error message: In is.na(data): is.na() applied to non-(list or vector) of type "closure"

- Warning error message: In mean.default(X): argument is not numeric or logical: returning NA

- Warning error message: In min/max(X): no non-missing arguments to min/max; returning Inf

- Warning message: In scan(file = file, what = what, sep = sep, quote = quote, dec = dec,: embedded nul(s) found in input

- Warning error message: invalid factor level, NA generated

- Warning error message: longer object length is not a multiple of shorter object length

- Warning error message: NAs introduced by coercion

- Warning error message: Removed X rows containing missing values

- Warning error message: the condition has length >1 and only the first element will be used

- Warning error message in as.POSIXct.POSIXlt(x): unknown timezone

- Warning error message in Ops.factor: not meaningful for factors

- Warning message in read.table: incomplete final line found by readTableHeader

- Warning error messages: In plot.window(...): nonfinite axis limits [GScale(-inf,1,1,.); log=1]

- Warning error messages: Removed 17 rows containing non-finite values (stat_bin).

MANIPULATION OF CONDITIONS IN R

In general, if we encounter unexpected errors while executing a program, we need an efficient and interactive way to debug the error and find out what happened. Try catch block is used for exception handling. The code (or set of statements) that can throw an exception is placed inside try block, and if the exception is raised, it is handled by the corresponding catch block. In this section, we will see various examples to understand how to use try catch for exception handling in R.

However, some errors are expected, where sometimes the models don't fit and throw an error. There are basically three ways to handle such conditions and errors in R.

- try(): helps us to continue executing the program even if an error occurs.

- tryCatch(): helps handle conditions and control what happens based on conditions.

- withCallingHandlers(): this is an alternative to tryCatch() that takes care of local handlers.

- try-catch-finally in R.

- Unlike other programming languages like Java, C++, and so on, try-catch-finally statements are used as functions in R. The main two conditions to handle in tryCatch() are "errors" and "warnings".

Let's understand try() using an example and its syntax.

We need to enclose unwanted commands in a try block. The commands passed inside are like function arguments. In case you have more than one statement, it is convenient to write a function with all those statements and call the function inside a try block.

In the following, there are various arguments that have a try block:

- expr: An R expression to try.

- silent: It should error message reporting be suppressed.

- outFile: Connection or character string naming the file to print to (via cat(*, file = outFile)); only used if silent is set to false, as the default.

Example:

```
> v<-list(1,2,4,'9',10, 0, 5)
> for (i in v) {
+       try(print(5/i))
+ }
[1] 5
[1] 2.5
```

```
[1] 1.25
Error in 5/i : non-numeric argument to binary
operator
[1] 0.5
[1] Inf
[1] 1
```

Using the try block, we can see that the code ran in all other cases, even after an error in one of the iterations. Now talk about tryCatch() using an example and its syntax.

The try-catch() function in R evaluates an exception-catching expression. The class of exception thrown by the standard stop() call is trial and error. The try-catch() allows users to handle errors. Using it you can do things like: if(error), then(this). In tryCatch(), there are two "conditions" that can be handled, such as 'warnings' and 'errors'. When writing each block of code, it is good to understand the execution state and scope.

Here is the syntax of tryCatch().

```
check_error = tryCatch({
    # expression
}, warning = function(w){
    # code that handles the warnings
}, error = function(e){
    # code that handles the errors
}, finally = function(f){
    # clean-up code
})
```

Example:

```
> # R program illustrating error handling
> # by using applying tryCatch
> tryCatch(
+
+       # Specifying expression
+       expr = {
+           1 + 1
+           print("Everything was fine.")
+       },
+       # Specifying error message
+       error = function(e){
+           print("There was an error message.")
```

```
+        },
+
+        warning = function(w){
+             print("There was a warning message.")
+        },
+
+        finally = {
+             print("finally Executed")
+        }
+ )
[1] "Everything was fine."
[1] "Finally Executed"
```

Another example:

```
> v<-list(1,2,4,'9',10, 0, 5)
> for (i in v) {
+        tryCatch(print(5/i),error=function(e){
+             print("Non conformabale arguments")
+        })
+ }
[1] 5
[1] 2.5
[1] 1.25
[1] "Non conformabale arguments"
[1] 0.5
[1] Inf
[1] 1
```

A try block doesn't let your code stop, but it doesn't provide any mechanism to handle an exception. Here, we can use a tryCatch block to handle the exception. We enclose the unwanted command in the tryCatch block and pass one more tryCatch parameter, error.

An error takes a function or instruction as input, and we can take any corrective action that needs to be taken in the event of an error in that function.

There are various parameters used in the syntax of tryCatch() given below:

- expression: It is to be evaluated.

- ...: It is a list of named expressions. An expression with the same name as the class of the exception thrown when the expression is evaluated.

- finally: It is an expression that is guaranteed to be called even if the expression throws an exception.

- envir: The envr in which the captured expression is to be evaluated.

withCallingHandlers() in R

In R, withCallingHandlers() function is a variant of tryCatch(). The difference is tryCatch() deals with exiting handlers, whereas withCallingHandlers() only deals with local handlers.

Example:

```
# R program illustrating error handling
# Evaluation of tryCatch
check <- function(expression){
> # R program illustrating error handling
> # Evaluation of tryCatch
> check <- function(expression){
+
+        withCallingHandlers(expression,
+
+                                warning = function(w){
+
message("warning:\n", w)
+                                },
+                                error = function(e){
+                                        message("error:\n",
e)
+                                },
+                                finally = {
+                                        message("Completed")
+                                })
+ }
>
> check({10/2})
Completed
[1] 5
> check({10/0})
Completed
[1] Inf
> check({10/'noe'})
Completed
```

```
error:
Error in 10/"noe": non-numeric argument to binary
operator
Error in 10/"noe" : non-numeric argument to binary
operator
```

PROCESSING CONDITIONS IN R PROGRAMMING

Decision processing (condition processing) is an important point in any programming language. Most use cases lead to positive or negative results where sometimes there is an option to check the status of more than one n number of options. In this section, we will discuss how condition processing works in R.

Sometimes there is a possibility that R code will go wrong and they can be communicated through errors such as warnings and messages. Fatal errors will raise stop() and force all execution to terminate. Errors are used when a function cannot continue. Warnings are generated using warning() and used to indicate potential problems when some elements of the vectorized input are invalid, such as log(−0:2). Messages are generated using message() that are used to provide informative output in a way that can be easily suppressed by the user.

CHANGING OF CONDITIONS PROGRAMMATICALLY

In the R language, there are three various tools available for programmatic handling of conditions, including errors. try() gives the ability to continue execution even if an error occurs.

Example:

```
# here we assign try() to do some variable
success <- try(100 + 200)
print(success)
failure <- try("100" + "200")
print(failure)
```

Here, we get to know the class of the variable success.

```
# here we assign try() to do some variable
success <- try(100 + 200)
print(success)
failure <- try("100" + "200")
print(failure)
```

```
# Class of success
class(success)
```

The output will be as follows.

```
Error in "100" + "200" : the non-numeric argument
to binary operator
[1] 300
[1] "Error in \"100\" + \"200\" : the non-numeric
argument to binary operator\n"
attr(,"class")
[1] "try-error"
attr(,"condition")
[1] "numeric"
```

Here, we get to know the class of the variable failure.

```
# here we assign try() to do some variable
success <- try(100 + 200)
print(success)
failure <- try("100" + "200")
print(failure)

# Class of failure
class(failure)
```

The output will be as follows.

```
Error in "100" + "200" : the non-numeric argument to
binary operator
[1] 300
[1] "Error in \"100\" + \"200\" : the non-numeric
argument to binary operator\n"
attr(,"class")
[1] "try-error"
attr(,"condition")
[1] "try-error"
```

When we put the code in a try block, the code runs, it will even error and for correct results it will be the last evaluated result, if it fails it will give a "try-error." tryCatch() function specifies a handler that controls what happens when a condition is signaled. The different actions can be taken for warnings, messages, and interrupts.

Example:

```
# Using tryCatch()
get_condition <- function(input code)
{
tryCatch(input code,
  error = function(c) "Unexpected error occurred",
  warning = function(c) "The warning message, but
still need to look into code",
  message = function(c) " The friendly message,
but take precautions"
             )
}

# Calling the function
get_condition(stop("!"))
get_condition(warning("?!"))
get_condition(message("?"))
get_condition(10000)
```

The output of the code is as follows:

```
[1] "Unexpected error occurred"
[1] "warning message, but\tstill need to look into
code"
[1] "friendly message, but\t\t\ttake precautions"
[1] 10000
```

withCallingHandlers() is an alternative to tryCatch(). It loads local drivers while tryCatch() registers existing drivers. It is useful for handling the message using the withCallingHandlers() method rather than try-Catch(), as this method will stop the program.

Example:

```
# Using tryCatch()
message_handler <- function(c) cat(" The important
message is caught!\n")
    tryCatch(message = message_handler,
    {
        message(" The first value printed ")
      message(" The second value too printed ")
    })
```

The output will be as follows.

```
The important message is caught!
```

Custom Signal Classes

A problem with error handling in R is the functions; just call stop() with a string. For example, "expected" errors (such as the model failing to converge for some input data sets) can be silently ignored, while unexpected errors (such as lack of free disk space) can be forwarded to the user.

Example:

```
# R program for illustration
# Custom signal classes

> condition <- function(subclass, message,
  call = sys.call(-1), ...) {
          structure(
          class = c(subclass, "condition"),
          list (message = message, call = call),
          ...
      )
    }

> is.condition <- function(x) inherits(x,
"condition")
> e <- condition(c("my_error", "error"),
               "An unexpected error occurred")
               stop(e) # Output as An unexpected
error occurred

#comment stop(s)
> w <- condition(c("my_warning", "warning"),
               "Appropriate warning!!!!")
> warning(w) # Output as appropriate warning!!!!
# the same as important message to note will be
printed

> m <- condition(c("my_message", "message"),
               "Important message to note!!!")
               message(m) # Output as an
important message to note!!!
```

The output will be as follows:

```
Error: Unexpected error occurred
Execution halted
```

DEBUGGING IN R PROGRAMMING

It is the process of cleaning the program code from errors so that it can run successfully.[1] When writing code, some errors or problems automatically appear after the code is compiled and are harder to diagnose. So the fix will take a lot of time and after several levels of calls.

Debugging in R is done using warnings, messages, and errors. This means debugging features. The various debugging functions are as follows:

- Editor breakpoint

- trace()

- browser()

- recover()

Editor Breakpoints

In RStudio, it can be added by clicking to the left of the row in RStudio Console or by pressing Shift + F9 with the cursor on the row. The breakpoint is the same as browser(), but it does not involve changing codes. Breakpoints are marked with a red circle on the left, indicating that debug mode will be entered on this line when the source is started.

Example:

```
> # Function 1
> new_function_1 <- function(a){
   a + 5
}
> # Function 2
> new_function_2 <- function(b) {
   new_function_1(b)
}
> # Calling function
> new_function_2(2)
> # Call traceback()
> traceback()
```

The traceback()

traceback() is used to provide all the information about how your function arrived at the error. It shows all functions called before an error occurs, called the "call stack" in many languages; R prefers call traceback.

Example:

```
# Function 1
function_1 <- function(a){
a + 5
}

# Function 2
function_2 <- function(b){
function_1(b)
}

# Calling error handler
options(error = traceback)
function_2(5)
Output:
[1] 10
```

browser() Function

The browser() function is embedded in the functions to open the interactive R debugger. It stops the execution of function() and you can explore the function with the environment itself. In debug mode, we can modify objects, view objects in the current environment, and also continue execution. The consoles confirm that you are in debug mode. There are some commands to follow:

- ls(): These are objects available in the current environment.

- print(): Used to evaluate objects.

- n: It is to examine another statement.

- s: It is to examine the next command by entering the function call.

- where: It is a stack trace printout.

- c: Is to exit the debugger and continue execution.

- C: This exits the debugger and returns to the R prompt.

Example:

```
# Calling recover
options(error = recover)

# Function 1
function_1 <- function(a){
a + 5
}

# Function 2
function_2 <- function(b) {
function_1(b)
}

# Calling function
function_2(45)
```

Output:

```
[1] 50
```

recover() Function

The recovery() command is used as an error handler and not as a direct command. In recovery(), R prints the entire call stack and lets you choose which function viewer to specify. A debug session will then start in the selected location.

Example:

```
# Calling recover
options(error = recover)

# Function 1
function_1 <- function(a){
  a + 5
}

# Function 2
function_2 <- function(b) {
 function_1(b)
}

# Calling function
function_2(30)
```

Output:

```
[1]  35
```

DEBUGGING TOOLS IN R

R provides various numbers of tools to help you debug your code.[2] The primary tools for debugging functions are as follows:

- traceback(): It prints the function call stack when an error occurs; does nothing if there is no error.

- debug(): It marks the function for "debug" mode, which allows you to step through the execution of the function line by line.

- browser(): It pauses execution of the function wherever it is called and switches the function to debug mode.

- trace(): It allows to inject debug code into a function at specific locations.

- recovery(): It allows you to modify the error behavior so that you can traverse the function call stack.

The traceback() function prints the function call stack after an error occurs. The function stack is the sequence of functions that were called before the error occurred. The traceback() shows you how many levels you were when the error occurred.

Example:

```
> mean (x)
Error in mean(x)  : the object 'x' not found
> traceback()
1: medium (x)
```

The traceback() function should be called immediately after an error occurs. Once another function is called, you lose the traceback.

Using debug()

The debug() function starts an interactive debugger (also known as a "browser" in R) for the function. Using a debugger, you should step through the R function one expression at a time to determine exactly where the error occurred. The debug() function takes as its first argument. Here is an example of debugging lm() function.

Using recover()

The recovery() function can be used to modify the error behavior of R when an error occurs; R prints an error message, exits the function, and returns you to your workspace to wait for further commands.

FILE HANDLING IN R

In the R programming language, we deal with large amounts of data by representing data in files.[3] We can perform certain data access operations such as creating files, reading files, renaming them, and so on.

Some different file operations in R are as follows:

- Creating files

- Writing to files

- Reading data from a file

- Check the current status of the file

- Renaming existing files

Creation of a File

This is the first operation that is performed when manipulating files. R allows us to escape creating files with each other, for example, by allowing us to create runtime files in a specific location.

Here is the syntax, file.create("file-name-with-extension").

Example:

```
> #here we returns Boolean value
> # and creating file
> if (file.create("R_Programming.txt")) {
+      print('Congrats! Your File Has been
created.')
+ } else {
+      print( " Unable to Create File)
+ }
[1] " Your File Has been created."
> # returns Boolean value
```

In this example, we used the file.create() function to create a text file called "R_Programming.txt." This method returns a Boolean value of true if runs successfully. Otherwise, it returns false. The code below will print "Your file has been created." After successfully creating the file, it will print "Unable to create file."

Writing to Files

Writing to files is data manipulation operation in R. It provides the write. table() function that allows data to be written with respect to a table format.

Here is the syntax, write.table(x=data, file="file-name-with-extension").

Parameters

- x: It represents the data we want to write.

- file: It indicates the file to be written.

Example:

```
> # here we are calling write method
> write.table (x = ToothGrowth [1:10, ],
"R_Programming.txt")
> data = read.table ("R_Programming.txt")
> print (data) # printing data
     len supp dose
1    4.2   VC  0.5
2   11.5   VC  0.5
3    7.3   VC  0.5
4    5.8   VC  0.5
5    6.4   VC  0.5
6   10.0   VC  0.5
7   11.2   VC  0.5
8   11.2   VC  0.5
9    5.2   VC  0.5
10   7.0   VC  0.5
```

We added some rows from the ToothGrowth dataset built-in to the file R_Programming.txt using the write.table() function and printing data on the console using the print() method.

Reading Data from a File

After writing the data to the file, we need to read the information from the file using a built-in function. We use the read.table() function to read the contents of the file that is passed as an argument.

Here is the syntax, read.table("file-name-to-read-with-extension").

Parameters

file: It takes the file name with an extension as a parameter.

Example:

```
> write.table(x = ToothGrowth[1:10, ],
"R_Programming.txt")
> # reading data from R_Programming.txt
> data = read.table("R_Programming.txt")
> print(data) # printing data
     len supp dose
1    4.2   VC  0.5
2   11.5   VC  0.5
3    7.3   VC  0.5
4    5.8   VC  0.5
5    6.4   VC  0.5
6   10.0   VC  0.5
7   11.2   VC  0.5
8   11.2   VC  0.5
9    5.2   VC  0.5
10   7.0   VC  0.5
```

We are using the read.table() method to read data from R_Programming. txt which contains the first 10 rows of the ToothGrowth dataset and print on the console through the print() method.

Check an Existing File

We can check the file if it exists or not within the current directory or on the mentioned path using the file.exists() function. We pass the file name, and if the file name is in existence, it returns TRUE. Otherwise, it returns FALSE.

Here is the syntax, file.exist("file-name").

Example:

```
> if (file.exists("R_Programming.txt")) {
+     print('Your File 'R_Programming.txt.txt'
Exist!')
+ } else {
+     print('Alas! File 'R_Programming.txt.txt' is
Unavailable')
+ }
[1] "Your File 'R_Programming.txt.txt' Exist!"
> # Does not exist
> if (file.exists("Another_File.txt")) {
+     print('Your File 'Another_File.txt' Exist!')
+ } else {
+     print('Alas! File 'Another_File.txt' is
Unavailable')
+ }
[1] "Alas! File 'Another_File.txt' is Unavailable"
```

CHAPTER SUMMARY

Here, we covered the basic of error and file handling, in which the user can practice along with examples; also you got some knowledge of types of function used in error handling like try(), tryCatch(), and more. In file handling, we basically have some common operations like creation, reading, deletion, and writing.

NOTES

1. Debugging in R – https://www.geeksforgeeks.org/debugging-in-r-program-ming/, accessed on October 4, 2022.
2. Debugging Tools in R – https://bookdown.org/rdpeng/rprogdatascience/debugging.html, accessed on October 5, 2022.
3. File Handling – https://www.educative.io/answers/how-does-file-handling-work-in-r-programming, accessed on October 6, 2022.

## CHAPTER 9

Graphics in R

IN THIS CHAPTER

➤ Introduction

➤ Data Set

➤ Bar Chart

➤ R Plotting

➤ R Line Chart and Graph

➤ R Scatter Plot

➤ and More

The previous chapter is related to all the errors occurring in R programming; we also use some debugging tools to understand the error. Another topic is file handling in which we briefly talked about the reading and writing operation. Now another topic is graphic in R.

INTRODUCTION GRAPHIC

The R language is mostly used for statistical and data analysis purposes for graphical representation of data in software. Tables and graphs are used in R to graphically represent this data. There are 100 of charts and graphs present in R programming, for example, barplot, dot chart, cloplot, histogram, boxplot, mosaic plot, pie chart, scatter graph, and so on.

DOI: 10.1201/9781003358480-9

Types of R Charts

- Barplot

- Boxplot

- Density Plot

- Heatmap

- Line Plot

- Pairs Plot

- Polygon Plot

- QQplot

- Scatter Plot

- Venn Diagram

- Pie Diagram or Pie Chart

- Histogram

Before chart, we will discuss the built-in data set in R so that we can use them to get the dummy value.

DATA SET

A data set is a set of data, often presented in a table. There is a popular built-in data set in R called "mtcars" (Motor Trend Car Road Tests) which is taken from Motor Trend US Magazine from 1974.

In the examples below (and in subsequent chapters), we will use the mtcars data set for statistical purposes. The mtcars data set is a built-in data set that contains measurements on 11 different attributes for 32 different cars.[1] We use the mtcars data set, for statistical purposes.

Example:

```
# Here we print the mtcars data set,
> mtcars
                   mpg cyl  disp  hp
Mazda RX4         21.0   6 160.0 110
Mazda RX4 Wag     21.0   6 160.0 110
Datsun 710        22.8   4 108.0  93
```

Hornet 4 Drive	21.4	6 258.0	110
Hornet Sportabout	18.7	8 360.0	175
Valiant	18.1	6 225.0	105
Duster 360	14.3	8 360.0	245
Merc 240D	24.4	4 146.7	62
Merc 230	22.8	4 140.8	95
Merc 280	19.2	6 167.6	123
Merc 280C	17.8	6 167.6	123
Merc 450SE	16.4	8 275.8	180
Merc 450SL	17.3	8 275.8	180
Merc 450SLC	15.2	8 275.8	180
Cadillac Fleetwood	10.4	8 472.0	205
Lincoln Continental	10.4	8 460.0	215
Chrysler Imperial	14.7	8 440.0	230
Fiat 128	32.4	4 78.7	66
Honda Civic	30.4	4 75.7	52
Toyota Corolla	33.9	4 71.1	65

Information-Related Data Set

You can use this mark (?) to get all the information about the mtcars data set.

Example:

```
# here we use the question mark to get information
about the data set.
?mtcars
```

The output of the code is as follows:

mtcars {datasets}   R Documentation

Motor Trend Car Road Tests

Description

The data was extracted from the 1974 *Motor Trend* US magazine, and comprises fuel consumption and 10 aspects of automobile design and performance for 32 automobiles (1973–74 models).

Usage

mtcars

Format

A data frame with 32 observations on 11 (numeric) variables.

[, 1] mpg Miles/(US) gallon
[, 2] cyl   Number of cylinders
[, 3] disp Displacement (cu.in.)

Information-related data set.

Get Information

You can use the dim() function to find the dimensions of the data set, the names() function to see the names of the variables.

Example:

```
> # here we create a variable of the mtcars data
set for better organization.
> Data_Cars <- mtcars
> # You can use dim() to find the dimension of the
data set
> dim(Data_Cars)
[1] 32 11
> # You can use names() to find the names of the
variables from the data set
> names(Data_Cars)
 [1] "mpg"   "cyl"   "disp" "hp"    "drat" "wt"
"qsec"
 [8] "vs"    "am"    "gear" "carb"
```

You can use the rownames() function to get the name of each and every row in the first column, which is the name of each car.

Example:

```
Data_of_cars <- mtcars
rownames(Data_of_cars)
```

The output of the following:

```
> Data_Cars <- mtcars
> rownames(Data_Cars)
 [1] "Mazda RX4"            "Mazda RX4 Wag"
 [3] "Datsun 710"           "Hornet 4 Drive"
 [5] "Hornet Sportabout"    "Valiant"
 [7] "Duster 360"           "Merc 240D"
 [9] "Merc 230"             "Merc 280"
[11] "Merc 280C"            "Merc 450SE"
[13] "Merc 450SL"           "Merc 450SLC"
[15] "Cadillac Fleetwood"   "Lincoln Continental"
[17] "Chrysler Imperial"    "Fiat 128"
```

```
[19]  "Honda Civic"          "Toyota Corolla"
[21]  "Toyota Corona"        "Dodge Challenger"
[23]  "AMC Javelin"          "Camaro Z28"
[25]  "Pontiac Firebird"     "Fiat X1-9"
[27]  "Porsche 914-2"        "Lotus Europa"
[29]  "Ford Pantera L"       "Ferrari Dino"
[31]  "Maserati Bora"        "Volvo 142E"
```

Here is a brief explanation of the variables from the mtcars data set:[2]

Variable name	Explanation
mpg	It defines as miles/(US) gallon
cyl	It defines as the number of cylinders
disp	It is a displacement
hp	It defines as gross horsepower
drat	It is a rear axle ratio
wt	It is the weight (1000 lbs)
qsec	It is 1/4 mile time
vs	It is engine (0 = V-shaped, 1 = straight)
am	It is transmission (0 = automatic, 1 = manual)
gear	It represents as a number of forward gears
carb	It represents as a number of carburettors

Print Variable Values

If you want to print all values possible that belong to a variable, access the data frame by using this $ sign, and the name of the variable (e.g., cyl [cylinders]):[3]

Example:

```
Data_of_cars <- mtcars
Data_of_cars$cyl
```

The output of the code is as follows:

```
> Data_Cars <- mtcars
> Data_Cars$cyl
 [1]  6 6 4 6 8 6 4 4 6 6 8 8 8 8 8 8 4 4 4 4 8 8 8 8 4 4
[28]  4 8 6 8 4
```

Sort Variable Values

You can sort the values using the sort() function:

Example:

```
> Data_Cars <- mtcars
> sort(Data_Cars$cyl)
 [1] 4 4 4 4 4 4 4 4 4 4 4 4 6 6 6 6 6 6 6 6 8 8 8
8 8 8 8 8
[28] 8 8 8 8 8
```

Analyzing the Data

Now we have some information about the data set and we can analyze it with some statistical numbers. For example, we can use the summary() method to get a statistical summary of the data.

Example:

```
Data_of_ Cars <- mtcars
summary(Data_of_Cars)
```

The summary() function always returns statistical numbers for each variable:

- Min

- The first quantile (percentile)

- Median

- Mean

- The third quantile (percentile)

- Max

Max Min

R has various different built-in math functions. For example, the min() and max() functions can be used to find the lowest or highest value in a set.

Example:

```
> Data_Cars <- mtcars
> max(Data_Cars$hp)
```

```
[1] 335
> min(Data_Cars$hp)
[1] 52
```

Mean, Median, and Mode

In statistics, there are often three values that interest us:

- Mean: It is the average value.

- Median: It is the middle value.

- Mode: It is the most common value.

Mean

To find the average value (mean) of a variable from the mtcars data set, find the sum of all values and divide the sum by the number of values.

Example:

```
> Data_Cars <- mtcars
> mean(Data_Cars$wt)
[1] 3.21725
```

Median

The median value is the value in the middle, after you have sorted all the values.

Example:

```
> Data_Cars <- mtcars
> median(Data_Cars$wt)
[1] 3.325
```

Mode

It is the value that only appears the most number of times. R doesn't have any function to calculate the mode. However, we can create our function to find it.

Example:

```
> Data_Cars <- mtcars
> names(sort(-table(Data_Cars$wt)))[1]
[1] "3.44"
```

Visualize the mtcars Data Set

We can create some plots to visualize the values in the data set.

You can use the hist() function to create a histogram of the values for a certain variable.

Example:

```
> hist(mtcars$mpg,
+        col='steelblue',
+        main='Histogram',
+        xlab='mpg',
+        ylab='Frequency')
```

To Create a Scatter Plot

We can use the plot() function to create a scatter plot of pairwise combination of variables.[4]

Example:

```
> #create scatterplot of mpg vs. wt
> plot(mtcars$mpg, mtcars$wt,
+        col='steelblue',
+        main='Scatterplot',
+        xlab='mpg',
+        ylab='wt',
+        pch=19)
```

Loading the mtcars Data Set in R Code

Since the mtcars data set is a built-in data set in R, you can load it by using the following command, such as data(mtcars)

We can see the data set by using the head() function.

```
> head(mtcars)
```

	mpg	cyl	disp	hp	drat
Mazda RX4	21.0	6	160	110	3.90
Mazda RX4 Wag	20.0	6	160	110	3.90
Datsun 710	22.8	5	108	93	3.85
Hornet 4 Drive	20.4	6	258	110	3.08
Hornet Sportabout	18.7	8	360	175	3.15
Valiant	18.1	6.1	225	105	2.76

```
                   wt   qsec vs am gear
```

Mazda RX4	2.620	16.46	0	1	4
Mazda RX4 Wag	2.775	17.02	0	1	4
Datsun 710	2.321	18.61	1	1	4
Hornet 4 Drive	3.205	19.44	1	0	3
Hornet Sportabout	3.440	17.02	0	0	3
Valiant	3.460	20.22	1	0	3

	carb
Mazda RX4	4
Mazda RX4 Wag	4
Datsun 710	1
Hornet 4 Drive	1
Hornet Sportabout	2
Valiant	1

Each type of graphic is illustrated with a basic code example. These codes are totally based on the following data.

BAR CHART

A bar chart presents data in rectangular bars with the length of the bar proportional to the value of the variable.[5] R uses the barplot() function to create bar graphs. R can draw both vertical and horizontal bars in a bar chart. In a bar chart, each bar can have a different color.

Create barplots using the barplot(height) function, where height is a vector or matrix. If height is a vector, the values determine the heights of the bars in the graph. If the height of the matrix and the option next to it is FALSE, then each column of the graph corresponds to a column of height, with the values in the column indicating the heights of the stacked "sub-columns." If the height of the matrix and next to it is TRUE, then the values in each column are next to other that are not on top of each other. Include the names.arg=(character vector) option to label the bars. Option horiz=TRUE to create a horizontal barplot.

Syntax

The basic syntax for creating a bar chart is:

```
barplot(H, xlab, ylab, main, names.arg,col)
```

The following is a description of the used parameters:

- H: It is a vector or matrix containing the numerical values used in the bar chart.

- xlab: It is the label used for the *x*-axis.

- ylab: It is the label used for the *y*-axis.

- main: It is the name of the bar graph.

- names.arg: This is the vector of names appearing under each bar.

- col: Used to color the bars in the chart.

Let's discuss various examples of R bar chart.

First example:

```
> # Here we have simple Bar Plot
> counts <- table(mtcars$gear)
> barplot(counts, main=" Distribution ",
    xlab=" Name of the xlab local variable")
```

The output of the code is as follows:

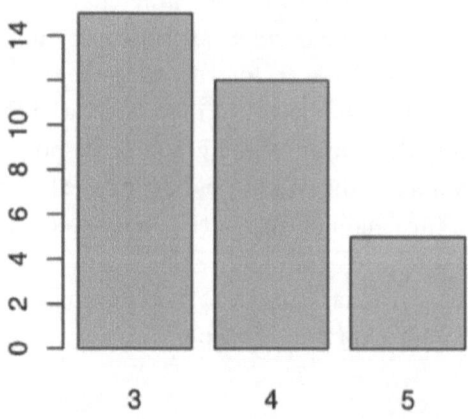

Bar chart 1.

Second example:

```
# here we define vector
> x <- c(7, 15, 23, 12, 44, 56, 32, 42,  88)
```

```
# Here is the output to be present as PNG file
> png(file = "bar_plot.png")

# plotting vector
> barplot(x, xlab = "xlab local variable",
        ylab = "Count", col = "white",
        col.axis = "darkblue",
        col.lab = "darkgreen")

# here we saving the file
> dev.off()
```

The output of the code is as follows:

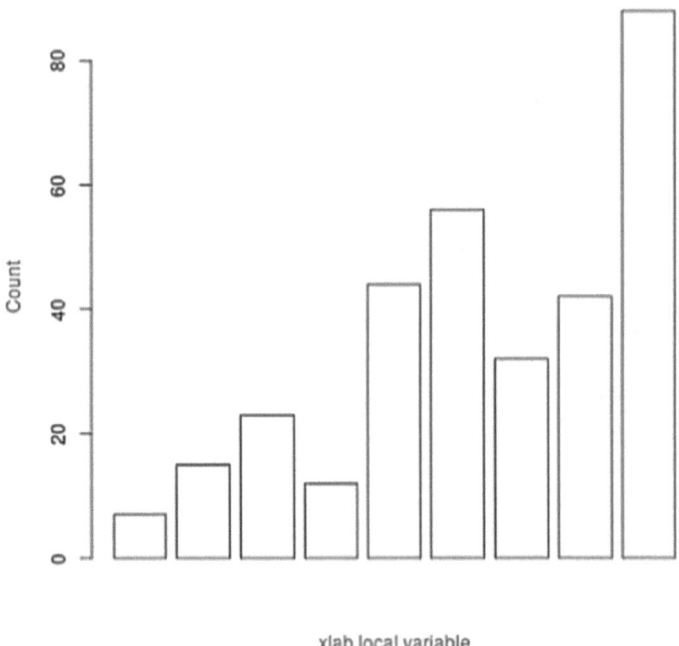

xlab local variable

Bar chart 2.

Third example:

```
# Here we have simple horizontal Bar Plot with
Added Labels
counts <- table(mtcars$gear)
```

```
barplot(counts, main="Car Distribution",
horiz=TRUE,
  names.arg=c(" Variable 1 ", " Variable 2 ",
"Variable 2"))
```

The output of the code is a follows:

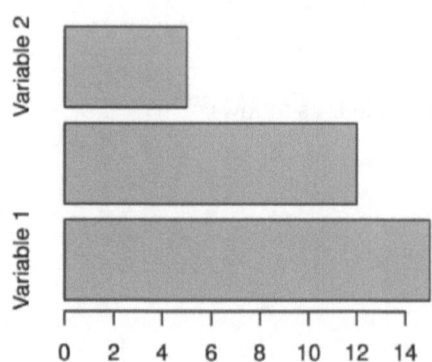

Bar chart 3.

Fourth example:

```
> # Create the data for the chart
> H <- c( 7, 12, 28, 3, 41, 10, 35)
> Months <- c( "March ", "April", "May", "Jun",
"Jul", "Aug", "Sep")

# Give the chart file a name
> png(file = "barchart_revenue.png")

# Plot the bar chart
> barplot(H,names.arg=M,xlab="Month",ylab="Revenue
",col="lightblue",
> main="Revenue chart",border="red")

# Save the file
> dev.off()
```

The output of the code is as follows:

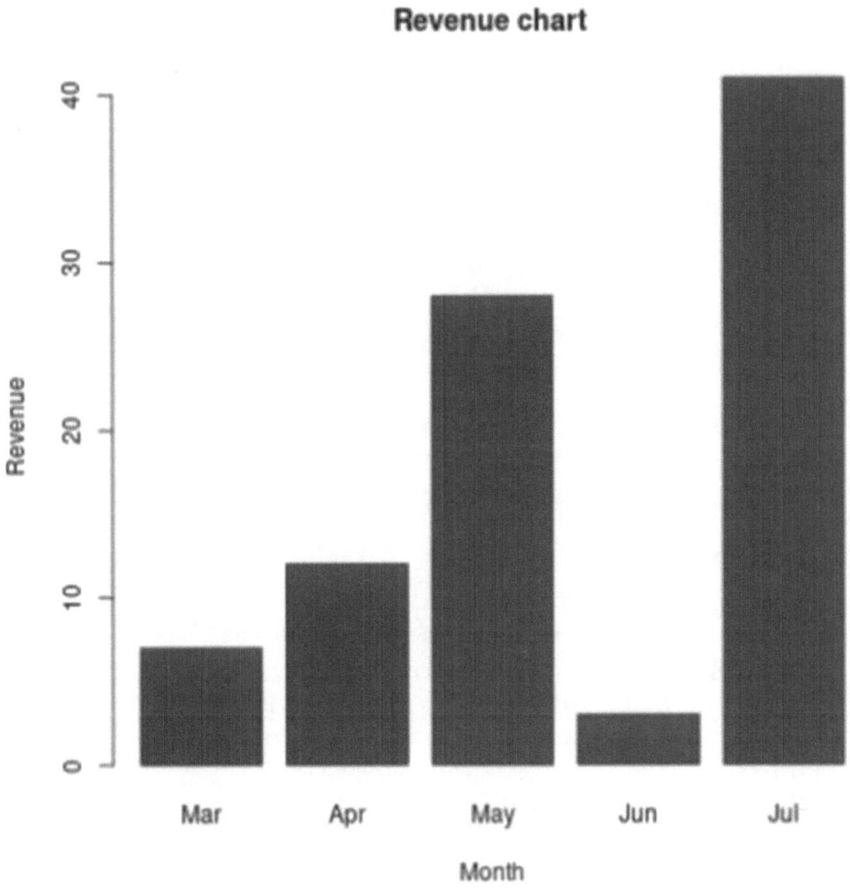

Revenue chart

Bar chart 4.

Fifth example:

```
> # Here we have stacked Bar Plot with Colors and
Legend
> counts <- table(mtcars$vs, mtcars$gear)
> barplot(counts, main=" Distribution   ",
  xlab=" xlab local variable ", col=c("lightblue",
"lightpink"),
  legend = rownames(counts))
```

The output of the code is as follows:

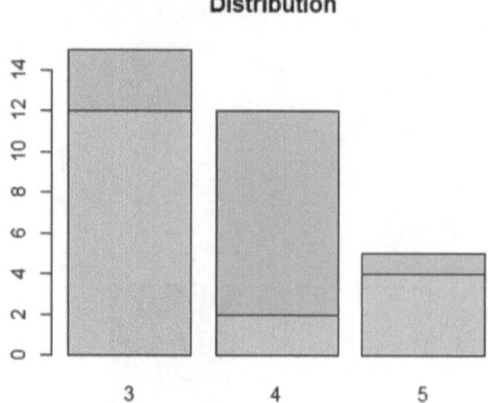

Bar chart 5.

Sixth example:

```
# Here we have grouped Bar Plot
counts <- table(mtcars$vs, mtcars$gear)
barplot(counts, main=" Distribution ",
   xlab=" xlab variable ", col=c("pink","red"),
   legend = rownames(counts), beside=TRUE)
```

The output of the code is as follows:

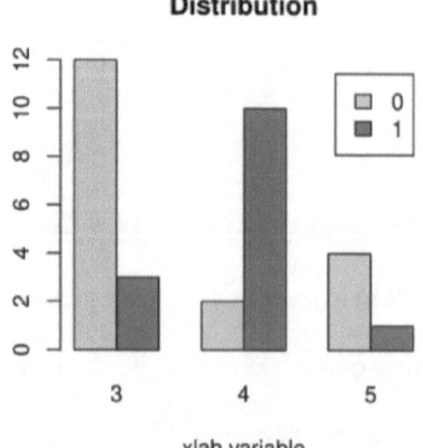

Bar chart 6.

Bar charts do not have to be based on counts or frequencies. You can create bar graphs that represent means, medians, standard deviations, and so on by using the aggregate() function and pass the results to the bar-plot() function.

By default, the categorical axis line is suppressed. To draw it, use the axis.lty=1 option.

For many lanes, the lane labels may begin to overlap. You can reduce the font size using the cex.names = option. Values less than one will reduce the size of the label. Additionally, you can use graphical parameters such as the following to help with text spacing.

```
# here we are fitting Labels
par(las=2) # make label text perpendicular to axis
par(mar=c(5,8,4,2)) # increase y-axis margin.

counts <- table(mtcars$gear)
barplot(counts, main=" Distribution ", horiz=TRUE,
names.arg=c(" 3 Variable ", "4 Variable", " 5
Variable"), cex.names=0.8)
```

The output of the code is as follows:

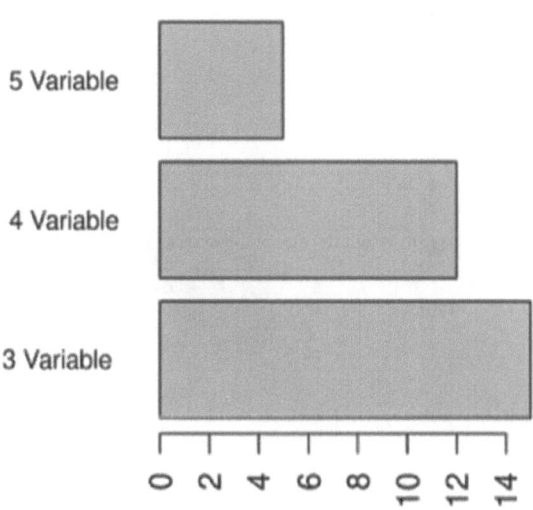

Bar chart 7.

R PLOTTING

There are various plotting shapes used in R programming, such as:[6]

- Plot

- Line

- Scatterplot

- Pie charts

- Bars

The plot() function is used to draw points (markers) in a diagram. It takes parameters for specifying points in the diagram. First specifies points on the x-axis. Second specifies points on the y-axis. At its simplest, you can use the plot() function to plot two numbers against each other.

#here we generate some sample data and plot graph.

```
> set.seed(1)
> # Here we generate sample data
> x <- rnorm(500)
> y <- x + rnorm(500)
> # Plot the data
> plot(x, y)
> # Equivalent using c bind
> M <- cbind(x, y)
> plot(M)
```

The output of the code is as follows:

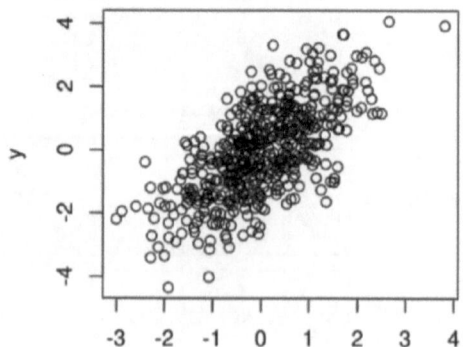

Plot chart 1.

With the plot function, you can create a various range of graphs, depending on the inputs. In the following table, we define all the available possibilities for the base R plotting function.

You can use the plot function to create a wide variety of graphs depending on the inputs. In the following table, we summarize all available options for the basic R plotting function.

There are some functions and arguments.

- plot(*x*, *y*): A scatter plot of the number vectors *x* and *y*.

- plot(factor): It plots a barplot of the factor.

- plot(factor, *y*): It plots a boxplot of a numeric vector and factor levels.

- plot(time_series): It plots a time series plot.

- plot(data_frame): It plots the correlation plot of all columns of the data frame (more than two columns).

- plot(date, y): It plots a vector based on a date.

- plot(function, lower, upper): It plots a function between the lower and maximum specified value.

R PLOT TYPE

You can also modify the plotter type using the type argument.[7] The choice of type will depend on the data you are plotting. In the following coding block, we will show the most popular types of graphs in R.

Example:

```
j <-  1 : 50
k <- j
par(mfrow = c(1, 4))
plot(j, k, type = "m", main = "type = 'm'")
plot(j, k, type = "p", main = "type = 'p'")
plot(j, k, type = "s", main = "type = 's")
par(mfrow = c(1, 1)
par(mfrow = c(1, 3))
plot(j, k, type = "i", main = "type = 'o'")
plot(j, k, type = "e", main = "type = 'b'")
plot(j, k, type = "d", main = "type = 'h'")
par(mfrow = c(1, 1))
```

There are various plot types used, as follows:

- p: It is point plot (default)
- l: It is line plot
- b: It is both (points and line)
- o: It is both (overplotted)
- s: It stairs plot
- h: It is histogram-like plot
- n: It is no plotting

R PLOT PCH

The pch argument allows you to modify the symbol of points in the graph. The main symbols can be selected with the numbers 1–25 as parameters. You can also change the symbol size with the cex argument and the symbol row width (except 15–18) with the lwd argument.

Example:

```
> r <- c(sapply(seq(5, 25, 5), function(i) rep(i,
5)))
> t <- rep(seq(22, 5, -5), 5)
> plot(r, t, pch = 1:22, cex = 3, yaxt = "n", xaxt
= "n",
        ann = FALSE, xlim = c(3, 27), lwd = 1:3)
text(r - 1.6, t, 1:25)
```

The output of the code is as follows:

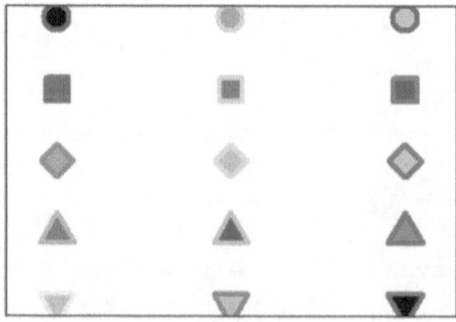

Plot chart 2.

Another example:

```
> r <- c(sapply(seq(5, 25, 5), function(i)
rep(i, 5)))
> t <- rep(seq(25, 5, -5), 5)

> plot(r, t, pch = 21:30, cex = 3, yaxt = "n",
xaxt = "n", lwd = 4,
     ann = FALSE, xlim = c(3, 27), bg = 1:30,
col = rainbow(25))
```

The output of the code is as follows:

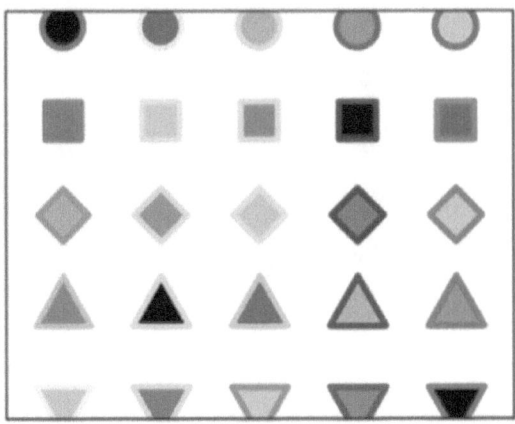

Plot chart 3.

In the following code, we show a simple example of how to customize one of these symbols.

Example:

```
x <- c(sapply(seq(5, 25, 5), function(i)
rep(i, 5)))
y <- rep(seq(25, 5, -5), 5)

plot(x, y, pch = 21:28, cex = 3, yaxt = "n", xaxt
= "n", lwd = 3,
     ann = FALSE, xlim = c(3, 27), bg = 1:28,
col = rainbow(25))

plot(x, y, pch = 21,
```

```
bg = "red",     # Fill color
col = "blue",   # Border color
cex = 3,        # Symbol size
lwd = 3)        # Border width
```

The output of the code is as follows:

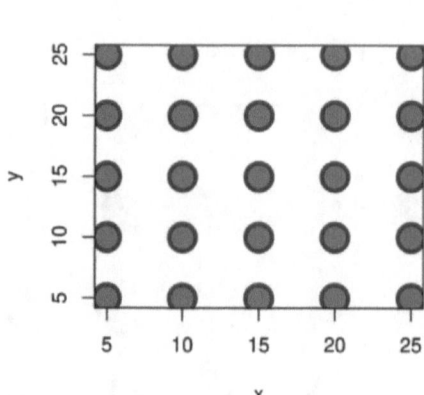

Plot chart 4.

You can use any character as a symbol. In fact, some character symbols can be selected using numbers 33–240 as the parameter of the pch argument.

```
# Custom symbols
x <- c(sapply(seq(5, 25, 5), function(i) rep(i, 5)))
y <- rep(seq(25, 5, -5), 5)
plot(1:5, 1:5, pch = c("☺", "♥", "✄", "✳", "✈"),
     col = c("green", 2:5), cex = 3,
     xlim = c(0, 5), ylim = c(0, 5))
```

Plot Title in R

You can add title to a plot with the main argument or the title function.[8]

```
plot(x, y, main = "This is a title")
# Here is the equivalent
plot(x, y)
title("My title")
```

Example:

```
x <- c(sapply(seq(5, 25, 5), function(i)
rep(i, 5)))
y <- rep(seq(25, 5, -5), 5)
plot(x, y, main = "My title")

# Equivalent
plot(x, y)
title("My title")
```

The output of the code is as follows:

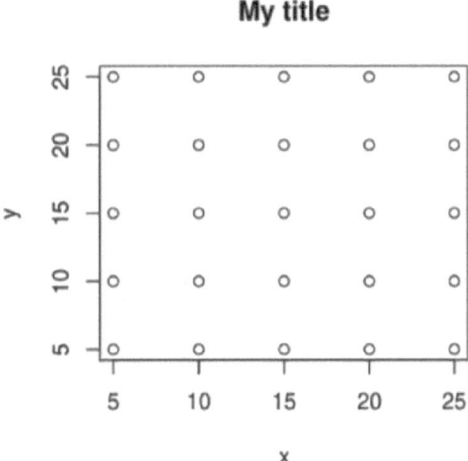

Plot chart 5.

SUBTITLE IN R PLOT

You can add a subtitle to a plot with the subargument that will display under the plot. It is to add a subtitle even if you don't specify a title.

Example:

```
x <- c(sapply(seq(5, 25, 5), function(i)
rep(i, 5)))
y <- rep(seq(25, 5, -5), 5)
# Here is the equivalent
plot(x, y)
title(main = "My title", sub = "My subtitle")
```

The output of the code is as follows:

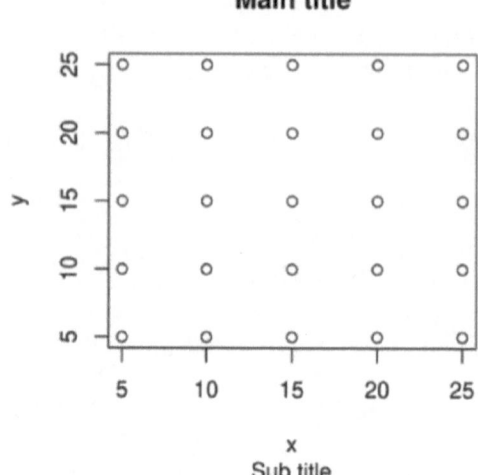

Plot chart 6.

AXIS IN R

In R plots, you can also modify the Y and X axis labels, add and change the axes tick, the size, and even set axis limits. By default, the R will use the vector names of plot as X and Y axes labels. However, you can also change them with the xlab and ylab arguments.

Example:

```
plot(x, y, xlab = "X label", ylab = "Y label")
```

If you want to delete the labels, you can set them to a blank string (" ") or set the ann argument to FALSE.

Example:

```
x <- c(sapply(seq(5, 25, 5), function(i)
rep(i, 5)))
y <- rep(seq(25, 5, -5), 5)
# here we delete labels
# make the string quotes empty
plot(x, y, xlab = "", ylab = "")
```

```
# Equivalent
plot(x, y, xlab = "X_label", ylab = "Y_label",
ann = FALSE)
```

The output of the code is as follows:

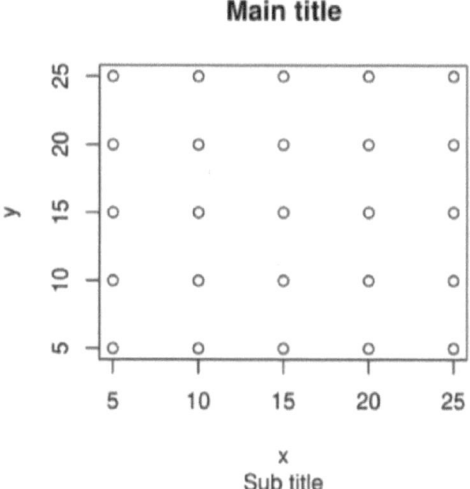

Plot chart 7.

R LINE CHART AND GRAPH

A line graph has a line that can connect all the points in the diagram.[9] If you want to create a line, use the plot() function and add a type parameter with a value of "l." A line graph is a graph that connects a series of points by drawing line segments between them. These points are ordered by one of their coordinate (usually x-coordinate) values. Line charts are typically used to identify trends in data. The plot() function is used to create a line graph. The basic syntax for creating a line plot in R is:

```
plot(v,type,col,xlab,ylab)
```

The following are descriptions of the used parameters:

- v: It is a vector containing numeric values.

- type: Has a value of "p" to draw only points, "l" to draw only lines, and "o" to draw both points and lines.

- xlab: It is a label used for the x-axis.

- ylab: It is label for the y-axis.

- main: It is used in the name of the chart.

- col: Used to color both points and lines.

Example:

```
# Create the data for the chart.
v <- c(7, 15, 22, 43, 10)

# Plot the bar chart.
plot(v, type = "o")
```

The output of the code is as follows:

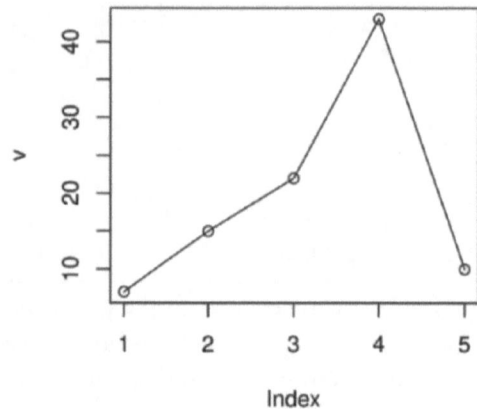

Line chart.

ADDING TITLE, COLOR, AND LABELS TO LINE CHARTS IN R

It takes all the parameters that are required to create a line graph by specifying a name for the graph and adding labels to the axes and we can add more features by adding more parameters with more colors to points and lines.

Example:

```
# Create the data for the chart.
v <- c(15, 25, 38, 13, 41)
```

```
# Plot the bar chart.
plot(v, type = "o", col = "green",
    xlab = "Month", ylab = " Article ",
    main = "Article Written chart")
```

The output of the code is as follows:

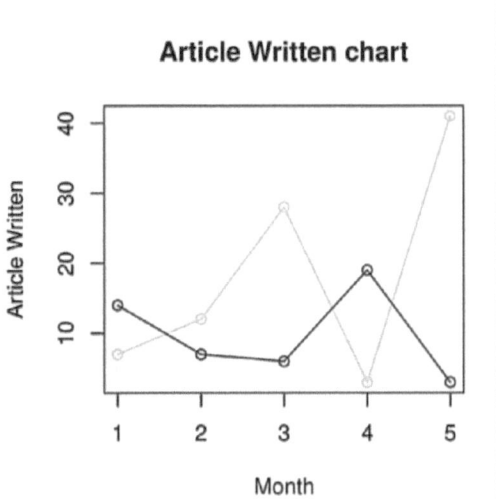

Line chart (single line).

MULTIPLE LINES IN A LINE CHART

You can add more than one line that can be drawn on a single graph using the lines() function. After drawing the first line, the lines() function can use another vector as input to draw the second line in the graph.

Example:

```
# Create the data for the chart.
v <- c(7,12,28,3,41)
t <- c(14,7,6,19,3)
# Plot the bar chart.
plot(v, type = "o", col = "green",
    xlab = "Month", ylab = "Article Written",
    main = "Article Written chart")
lines(t, type = "o", col = "blue")
```

The output of the code is as follows:

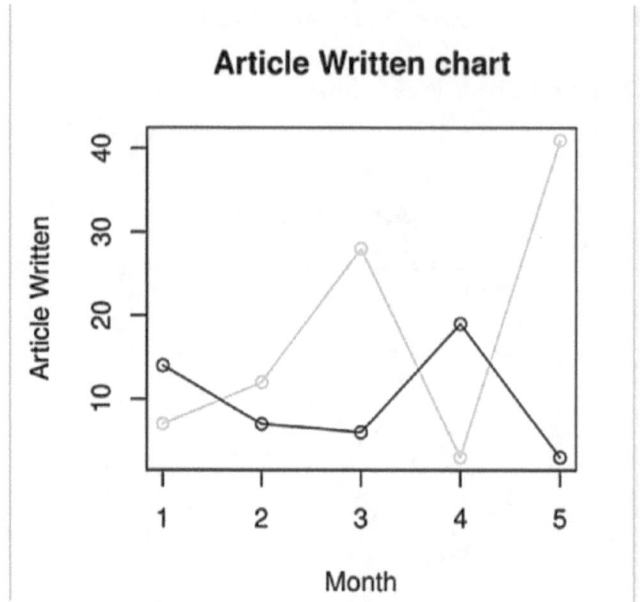

Line chart 2 (multiple line).

There are various properties of line graph.

- Color

- Width

- Styles

Let's discuss each with example.

Color

The line color is black by default, and to change the color, use the col parameter.[10]

```
# Plot numbers from 1 to 12 and draw a green line
> plot( 1:12, type="l", col="green")
```

The output of the code is as follows:

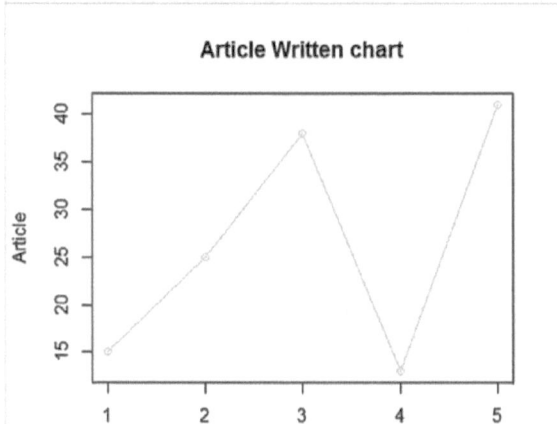

Line chart.

Width

If you want to change the width of the line, then use the lwd parameter (1 is default, while 0.05 means 50% smaller, and 2 means 100% larger).

Example:

```
# Plot numbers from 1 to 12 and draws a thick
line.
plot(1:12, type="l", lwd=5)
```

The output of the code is as follows:

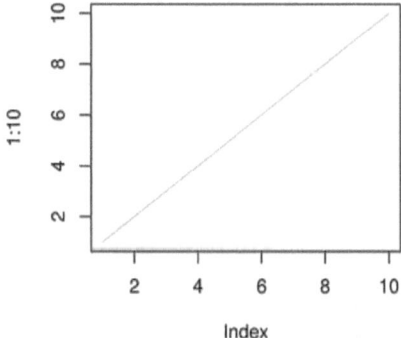

Line chart.

Style

The line is solid by default and you can use the lty parameter with a value from 0 to 6 only to specify the line format, where lty=3 will display a dotted line instead of a solid line.

```
# Plot numbers from 1 to 12 and draw a thick dotted
line
# code is here
plot(1:12, type="l", lwd=5, lty=5)
```

The output of the code is as follows:

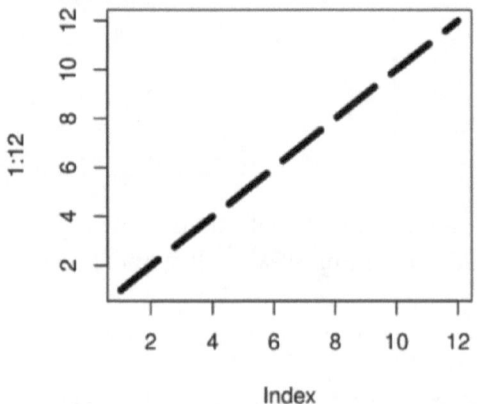

Line chart.

The following parameter values for lty are:

- 0: It removes the line
- 1: It displays a solid line
- 2: It displays a dashed line
- 3: It displays a dotted line
- 4: It displays a "dot dashed" line
- 5: It displays a "long dashed" line
- 6: It displays a "two dashed" line

MULTIPLE LINES

To display more than single line in a graph, use the plot() function together with the lines() function.

Example:

```
> line1 <- c(1, 3, 3, 5, 6, 8)
> line2 <- c(2, 5, 7, 8, 9, 10)
> plot(line1, type = "l", col = "blue")
> lines(line2, type="l", col = "red")
```

The output of the code is as follows:

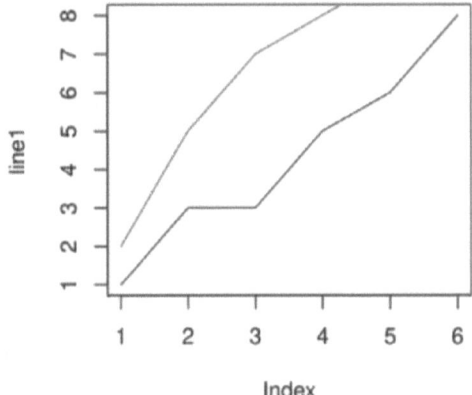

Multiple line.

R SCATTER PLOT

A "scatter plot" is a type of graph used to show the relationship between two numerical variables and plots one point for each observation.[11] Scatter plots show many points plotted on a Cartesian plane. Each and every point represents the values of two variables. One variable is selected on the horizontal axis and the other on the vertical axis.

It needs two vectors of equal length, one for the x-axis (horizontal) and one for the y-axis (vertical). The simple scatter plot is created using the plot() function. The basic syntax for creating scatterplot is:

```
plot(x, y, main, xlab, ylab, xlim, ylim, axes)
```

Following is the description of the parameters used:

- x: It is the data set whose values are the horizontal coordinates.

- y: It is the data set whose values are the vertical coordinates.

- main: It is the tile of the graph.

- xlab: It is the label in the horizontal axis.

- ylab: It is the label in the vertical axis.

- ylim: It is the limit of the values of y used for plotting.

- xlim: It is the limit of the values of x used for plotting.

- axes: These indicateswhether both axes should be drawn on the plot.

Here we have a simple example of plot below:

```
set.seed(12)
n <- 100
x <- runif(n)
eps <- rnorm(n, 0, 0.25)
y <- 2 + 3 * x^2 + eps
# In order to plot the observations that you type:
plot(x, y, pch = 19, col = "black")
plot(y ~ x, pch = 19, col = "black") # Equivalent
```

The output of the code is as follows:

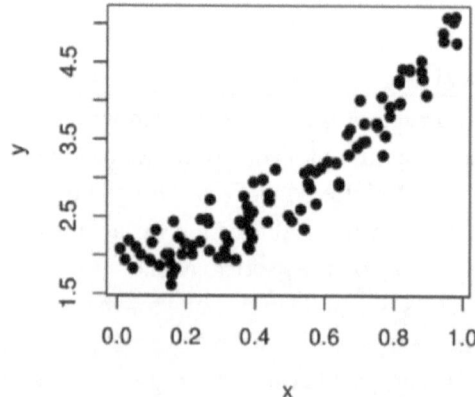

Scatter plot.

Creating the Scatterplot

We create a scatterplot graph for the relation between wt(weight) and mpg(miles per gallon). The example is given below.

```
# Get the input values.
input <- mtcars[,c('wt','mpg')]
# Plot the chart of cars with weight between 3 and 5
and mileage between 15 and 30.
plot(x = input$wt,y = input$mpg,
    xlab = "Weight",
    ylab = "Milage",
    xlim = c(3, 8),
    ylim = c(10, 40),
    main = " CarWeight vs Milage "
)
# Save the file.
dev.off()
```

The output of the code is as follows:

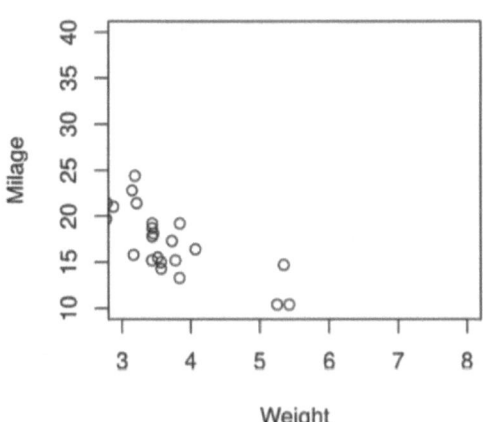

Scatter plot.

Scatter Matrices

If we have more than one variable and we want to find the correlation between one variable and the rest, we use the variance matrix. We use the pairs() function to create scatterplot matrices.

The basic syntax for creating simple scatterplot matrices in R is as follows:

```
pair (formula, data)
```

The following are descriptions of the used parameters.

- a formula: It represents a series of variables used in pairs.

- data: It represents the data set from which the variables will be taken.

PIE CHARTS

The R programming language has many libraries for creating tables and graphs. A pie chart is a representation of values as slices of a circle with different colors. The slices are labeled and the numbers corresponding to each slice are also given in the graph.

In R, a pie chart is created using the pie() function that only takes positive numbers as a vector input.[12] The other parameters are used to control labels, color, title, and so on.

The basic syntax for creating a pie chart using R is as follows:

```
pie (x, labels, radius, main, column, clockwise)
```

The following is the description of the parameters:

- x is a vector containing the numerical values used in the pie chart.

- Labels are used to describe slices.

- It specifies the radius of the circle of the pie chart (a value between −1 and +1).

- The main indicates the name of the chart.

- The col indicates the color palette.

- Clockwise is a Boolean indicating whether slices are drawn clockwise or counterclockwise.

Here is the example of pie chart.

```
# Create data for the graph.
x <- c(21, 62, 10, 53)
labels <- c(" UK ", " NY ", " NJ ", "India")0
# Plot the chart.
pie(x,labels)
```

The output of the code is as follows:

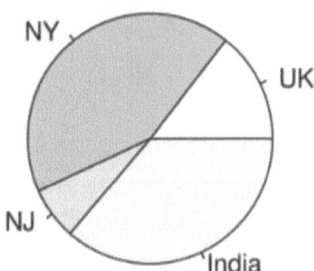

Pie chart.

Pie Chart Title and Colors

We can extend the properties of the graph by adding additional parameters to the function. We will use the main parameter to add the name of the graph, and the next parameter is col, which will use the rainbow color palette when drawing the graph. The length of the palette should be the same as the number of values we have for the chart. That's why we use length (x).

Example:

```
# Create data for the graph.
x <- c(11, 22, 50, 33)
labels <- c(" UK ", " NY ", " NJ ", "India")

# Plot the chart with title & rainbow color
pallet.
pie(x, labels, main = "Pie chart", col =
rainbow(length(x)))
```

The output of the code is as follows:

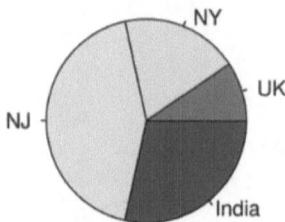

Pie chart

Pie chart.

Pie charts are created using the pie(x, labels=) function, where x is a non-negative numeric vector denoting the area of each slice and labels = denotes a character vector of slice names.

Example:

```
# Pie Chart with Percentages
slices <- c(10, 12, 4, 16, 8)
lbls <-  c(" UK ", " NY ", " NJ ", "India",
"France")
pct <- round(slices/sum(slices)*100)
lbls <- paste(lbls, pct) # add percents to labels
lbls <- paste(lbls,"%",sep="") # add % to labels
pie(slice, labels = lbls,
col=rainbow(length(lbls)),
    main=" Chart of Countries")
```

3D Pie Chart

The pie3D( ) function in the plotrix package provides 3D exploded pie charts. The 3D pie charts are not recommended for use, but if you really want to create, you can use pie 3D from plotrix package. The default 3D pie chart looks like the following.

Now first install the package named plotrix.

```
> install.packages("plotrix")
WARNING: Rtools is always required to build R packages
but is not current installed. Please download and
installed the appropriate version of Rtools before
moving,
```

```
https://cran.rstudio.com/bin/windows/Rtools/
Installing package into 'C:/Users/Dell/AppData/
Local/R/win-library/4.2'
(as 'lib' is unspecified)
trying URL 'https::/cran.rstudio.com/bin/windows/
contrib/4.2/plotrix_3.8-2.zip'
Content type 'application/zip' length 1138080 bytes
(1.1 MB)
downloaded 1.1 MB
package 'plotrix' successfully unpacked and MD5 sums
checked
The downloaded binary packages are in
        C:\Users\Dell\AppData\Local\Temp\RtmpExUU9C\
downloaded_packages
```

Now run the below code in the R console:

```
library(plotrix)
data <- c(19, 21, 54, 12, 36, 12)
pie3D(data)
```

   # 3D Exploded Pie Chart

```
library(plotrix)
slices <- c(11,  12, 4, 16, 8)
lbls <- c("US", "UK", "UEA", "Germany", "India","
France")
pie3D(slices, labels=lbls, explode=0.1,
   main="Pie Chart of Countries ")
```

Pie Chart Height in R

Similarly, you can also change the height of the pie with height. Default value is 0.1.

```
# install.packages("plotrix")
library(plotrix)
data <- c(19, 21, 54, 12, 36, 12)
pie3D(data,
      height = 0.2)
```

Pie Chart Angle in R

You can change the viewing angle with theta. The default value is pi/6.

Example:

```
# install.packages("plotrix")
library(plotrix)
data <- c(19, 21, 54, 12, 36, 12)
pie3D(data,  theta = 1.5)
```

BOXPLOT

Boxplots are a measure of how well the data in a data set is distributed.[13] It divides the data set into three quartiles. Its graph represents the minimum, maximum, median, first quartile, and third quartile in a data set. It is useful in comparing the distribution of data between data sets by drawing boxplots for each.

The boxplots are created by using the boxplot() function. The syntax to create a boxplot in R is as follows:

```
boxplot(x, data, notch, varwidth, names, main)
```

The following are descriptions of the used parameters:

- x: It is a vector or formula.

- data: It is a data frame.

- notch: It is a Boolean value. Set to TRUE to draw a notch.

- varwidth: It is a Boolean value. Set to true to render a frame width proportional to the sample size.

- names: It is the group labels that will be printed below each boxplot.

- main: It is used to indicate the name of the chart.

Histograms and Density Plots in R

A histogram is a graphical representation that only organizes a group of data points into user-specified ranges and approximates the distribution of numerical data.[14] In R, a histogram is created using the hist() function.

Syntax:

```
hist(v, main, xlab, xlim, ylim, break, col,
border)
```

Parameters:

- v: It is a vector containing the numerical values used in the histogram.

- main: Shows the chart name.

- col: Used to set the color of the bars.

- border: Use to set the border color of each bar.

- xlab: Used to describe the x-axis.

- xlim: Used to specify the range of values on the x-axis.

- ylim: Used to specify the range of values on the y-axis.

- breaks: Used to indicate the width of each bar.

Example:

```
> v <- c(5, 9, 13, 12, 40, 30, 49, 36, 3, 2, 18,
27,72,14)
> hist(v, xlab = "Weight", col = "green", border =
"black")
```

The output of the code is as follows:

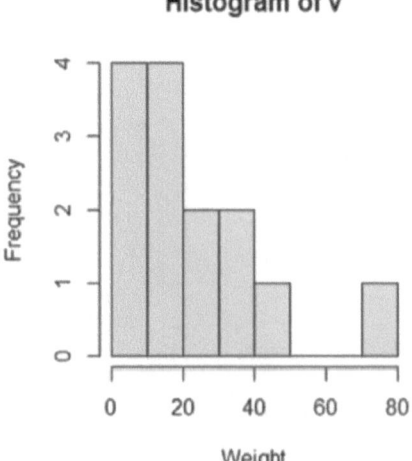

Histogram of V.

Density Plot

A density plot is a representation of the distribution of a numerical variable that uses a kernel density estimate to display the variable's probability density function. We use the density() function to help calculate kernel density estimates.

Syntax:

```
density(x)
```

Parameters: x: It is the data from which the estimate is to be computed.

A density plot is a representation of the distribution of a numerical variable that uses a kernel density estimate to display the variable's probability density function. We use the density() function to help calculate kernel density estimates.

```
> hist(beaver1$temp,
       col="green",
       border="black",
       prob = TRUE,
       xlab = "temp",
       main = "GFG")

> lines(density(beaver1$temp),
        lwd = 2,
        col = "chocolate3")
```

The output of the code is as follows:

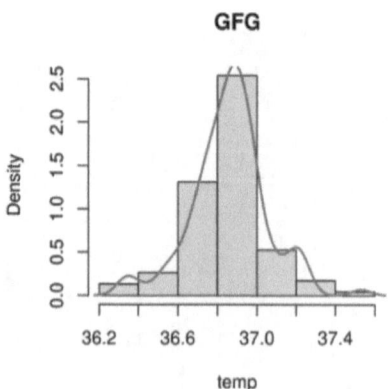

Density.

HEATMAP

A heatmap is defined as the graphical representation of data set where the individual values that contained in the matrix are represented as colors.

Using the Heatmap()

The heatmap() is natively provided in R. It produces a high-quality matrix and offers statistical tools to normalize input data, run a clustering algorithm, and visualize the result using dendrograms.

There are three ways to create an interactive heatmap from R:

- plotly: It allows to turn any heatmap created with ggplot2 into an interactive one.

- d3heatmap: It is a package that uses the same syntax as the base R heatmap() function to create an interactive version.

- heatmaply: The most flexible option that allows many different kinds of customization.

PARIS PLOT

In base R, you can create a pairwise correlation plot with the pairs function.[15] Note: This is the same as plotting a numeric data frame with plot.

Example:

```
#Numeric variables
df <- iris[1:4]
pairs(df)
# Equivalent to:
pairs(~ Sepal.Length + Sepal.Width +
        Petal.Length + Petal.Width, data = df)
# Equivalent to:
with(df, pairs(~ Sepal.Length + Sepal.Width +
                Petal.Length + Petal.Width))
```

The output of the code is as follows:

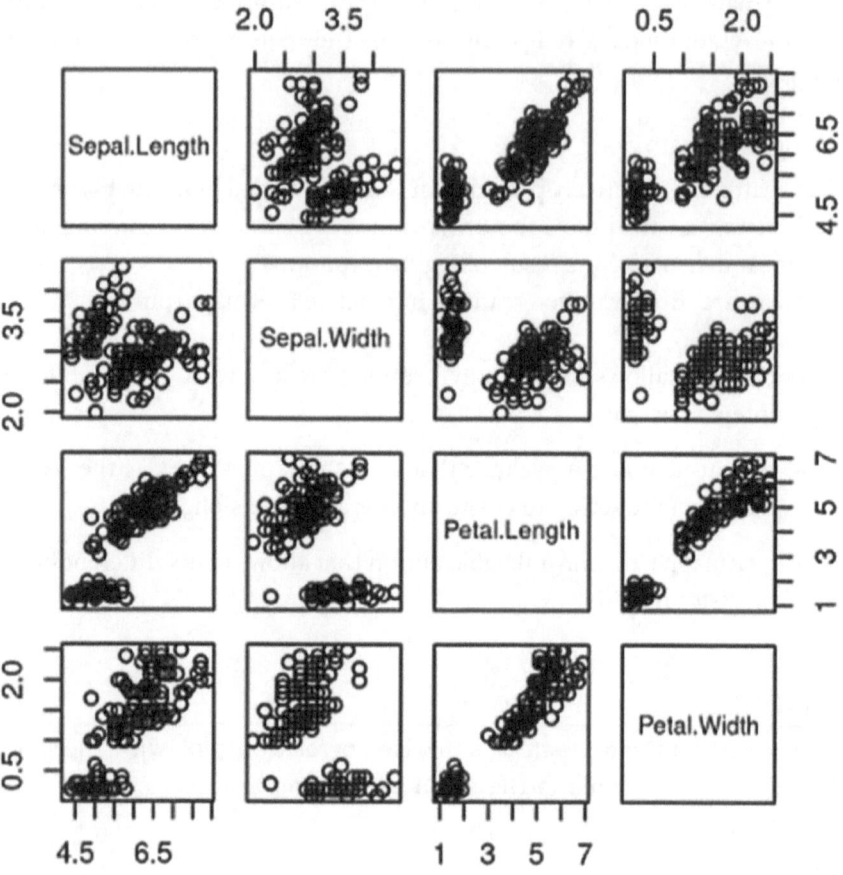

Paris plot.

VENN DIAGRAM

A Venn diagram shows all logical relationships between several sets of data. This section explains how to create it using R and the VennDiagram package, providing reproducible code.

The VennDiagram package allows you to create Venn diagrams, thanks to its venn.diagram() function. It takes a list of vectors as input. Each vector provides words. Now install the VennDiagram library and run this code.

```
# Load library
library(VennDiagram)
# Generate 3 sets of 100 words
```

```
set_1 <- paste(rep(" word_ ",  100),  sample(c(1 :
1000),   200,  replace=F),  sep="")
set_2 <- paste(rep(" word_ ",  100),
sample(c(1:1000),   200,  replace=F),  sep="")
set_3 <- paste(rep(" word_ ",  100),
sample(c(1:1000),   200,  replace=F),  sep="")
# Chart
venn.diagram(
  x = list(set_1, set_2, set_3),
  category.names = c("Set_1",  "Set_2 ",  "Set 3"),
  filename = '#14_venn_diagramm.png',
  output=TRUE
)
```

CHAPTER SUMMARY

Here we learned graphic in R, also its type of pie chart, box plot, histogram, Paris plot, and more. You can run the various coding examples to achieve the graphic in R.

NOTES

1. Dataset in R – https://www.statology.org/mtcars-dataset-r/#:~:text=The%20 mtcars%20dataset%20is%20a,the%20mtcars%20dataset%20in%20R, accessed on October 6, 2022.
2. Mtcars Data set – https://www.w3schools.com/r/r_stat_data_set.asp, accessed on October 6, 2022.
3. Data Set – https://www.w3schools.com/r/r_stat_data_set.asp, accessed on October 6, 2022.
4. Data Set – https://www.statology.org/mtcars-dataset-r/#:~:text=The%20mtcars %20dataset%20is%20a,the%20mtcars%20dataset%20in%20R, accessed on October 6, 2022.
5. Bar Chart – https://www.statmethods.net/graphs/bar.html, accessed on October 5, 2022.
6. Plotting – https://www.w3schools.com/r/r_graph_plot.asp, accessed on October 5, 2022.
7. Plot Type – https://r-coder.com/plot-r/#Plot_function_in_R, accessed on October 5, 2022.
8. Title Plot – https://r-coder.com/plot-r/#Plot_function_in_R, accessed on October 5, 2022.
9. Line Graph – https://www.tutorialspoint.com/r/r_line_graphs.htm, accessed on October 5, 2022.
10. Line Graph – https://www.w3schools.com/r/r_graph_line.asp, accessed on October 5, 2022.
11. Scatter Plot – https://www.tutorialspoint.com/r/r_scatterplots.htm, accessed on October 5, 2022.

12. Pie Chart – https://www.tutorialspoint.com/r/r_pie_charts.htm, accessed on October 6, 2022.

13. Box Plot – https://www.tutorialspoint.com/r/r_boxplots.htm, accessed on October 6, 2022.

14. Histogram – https://www.geeksforgeeks.org/histograms-and-density-plots-in-r/#:~:text=A%20density%20plot%20is%20a,to%20compute%20kernel%20density%20estimates, accessed on October 6, 2022.

15. Paris Plot – https://r-charts.com/correlation/pairs/, accessed on October 6, 2022.

Appraisal

The R Project for statistical computing, or simply R, is a free software environment for statistical graphics. It is also a programming language that is widely used as statisticians and data miners for statistical software development and data analysis. Over the past few years, they have been joined by businesses that have discovered R's potential as well as technology vendors that offer R support or R-based products.

Although there are other programming languages for statistical processing, R has become the de-facto statistical routine language and offers a package repository of over 6400 packages for solving problems. It also offers versatile and powerful rendering. It also has the advantage of treating tabular and multivariate data as a labeled, indexed series of observations. This is a game changer from typical software that only does two-dimensional layouts like Excel.

There are a number of integrated development environments (IDEs) that you can use to write R code, including Visual Studio for R, Eclipse, R Console, and RStudio, among others. You can use a plain text editor. However, we will be using RStudio for the exercises in this book. RStudio is a free, open-source IDE for R. It includes a console, a syntax-highlighting editor that supports code execution, as well as tools for plotting, debugging, and workspace management.

It is available in open-source commercial editions and runs on a desktop (such as Windows, Mac, and Linux) or in a web browser connected to RStudio Server or Server Pro (Debian/Ubuntu system, RedHat/CentOS, and SUSE Linux). Although there are many options for R development, we will use RStudio for the exercises in this book. You can get more information about RStudio at "https://www.rstudio.com/products/rstudio/." We have covered much of the user interface of RStudio in Chapter 2. There are various panes available in RStudio interface such as Files Pane, Plots Pane, Packages Pane, Help Pane, Viewer Pane, Environment Pane, History

DOI: 10.1201/9781003358480-10

Pane, Connections Pane, Source Pane, Console Pane, and Terminal Pane. The programming features are R packages and distributes' computing.

There are various detailed topics in R programming such as R Sessions and Functions, Basic Math, Variables, Data Types, Vectors, Advanced Data Structures, Data Frames, Lists, Matrices, Arrays, and Classes.

CAREER IN R

R has become for the data science and statistics. It is the most popular analytical tool. R is an open-source language intended for developers and programmers from all over the world. It is constantly expanding and people from all over the world contribute to its development.

R programmers can avail different types of jobs in the data science industry. Since there is a shortage of data scientists, both novice and professional programmers can enter the data science industry.

Some of the areas where the R applicants are most in demand are as follows:

- Financial sectors

- Banks

- Health-care organizations

- Manufacturing companies

- Academic

- Government departments

Companies like Google, Facebook, Accenture, MuSigma, and others are adopting the R platform. Some of the positions available for the programmers are as follows:

- Data scientist expert

- Business analyst expert

- Data analyst expert

- Data visualization expert

- Quantitative analyst expert

PROPERTIES OF R

R is a programming language with the environment for statistical analysis, graphical representation, and reporting. The following are important features of R:

- R is a well-developed, simple, and efficient programming language that includes conditionals, loops, user-defined recursive functions, and input and output devices.

- R has efficient data processing and storage facilities.

- R provides a set of operators for calculations on lists, arrays, vectors, and matrices.

- R provides a large, coherent, and integrated collection of data analysis tools.

- R provides graphical means for data analysis and displays either directly on the computer or printed on paper.

Index